Praise for Far

"A brilliant writer." — James Herriot

"Like Twain, Mowat perfectly captures the essence of a more innocent era, as seen through a young boy's eyes — and heart. And what a heart! And what a writer!"
— *San Francisco Chronicle*

"One of the finest nature writers on the planet."
— *Phoenix Gazette*

"Mr. Mowat did not pursue his alleged talent as a pornographer. He stuck to wildlife and oddball explorations and spirited comedy and is still at it, to the good fortune of his readers."
— *The Atlantic Monthly*

"Through both the notes and poems he wrote as a teenager and his own present-day descriptive powers, one is led to the writer's personal Parnassus — its groves filled with farmhouses and maple trees, its muse covered in feathers and fur."
— *New York Times Book Review*

"A warm and charming memoir about a family wise enough to recognize the preciousness of their life and surroundings."
— *London* (Ontario) *Free Press*

BOOKS BY FARLEY MOWAT

PEOPLE OF THE DEER • 1952

THE REGIMENT • 1955

LOST IN THE BARRENS • 1956

THE DOG WHO WOULDN'T BE • 1957

THE GREY SEAS UNDER • 1958

COPPERMINE JOURNEY • 1958

THE DESPERATE PEOPLE • 1959

ORDEAL BY ICE • 1960

OWLS IN THE FAMILY • 1961

THE SERPENT'S COIL • 1961

THE BLACK JOKE • 1962

NEVER CRY WOLF • 1963

WESTVIKING • 1965

THE CURSE OF THE VIKING GRAVE • 1966

CANADA NORTH • 1967

THE POLAR PASSION • 1967

THE ROCK WITHIN THE SEA • 1968

THE BOAT WHO WOULDN'T FLOAT • 1969

THE SIBERIANS (SIBIR) • 1970

A WHALE FOR THE KILLING • 1972

TUNDRA • 1973

WAKE OF THE GREAT SEALERS • 1973

THE SNOW WALKER • 1975

CANADA NORTH NOW • 1976

AND NO BIRDS SANG • 1979

THE WORLD OF FARLEY MOWAT • 1980
(edited by Peter Davison)

SEA OF SLAUGHTER • 1984

MY DISCOVERY OF AMERICA • 1985

WOMAN OF THE MISTS (VIRUNGA) • 1987

THE NEW FOUNDE LAND • 1989

RESCUE THE EARTH • 1990

MY FATHER'S SON • 1992

BORN NAKED

Farley Mowat

A PETER DAVISON BOOK

Houghton Mifflin Company

The photographs

p. viii – Farley Mowat as a Bay of Quinte Tadpole, 1922.

p. 1 – Angus Mowat and Helen Thomson, 1918.

p. 14 – Henry, the Mowat's first car and Angus's beehives, c. 1922.

p. 31 – Farley on the bridge of the tugboat *M. Sicken,* 1927.

p. 42 – The Mowat family boat, *Stout Fella,* c. 1928.

p. 60 – Angus yachting on Lake Erie, 1932.

p. 75 – Farley with his birdwatching binoculars en route to Saskatchewan, 1933.

p. 87 – Angus atop *Rolling Home,* aka *Angus's Ark,* 1936.

p. 103 – Farley with Mutt, taken in Saskatoon, 1935.

p. 116 – Farley and Helen at Caribou Gorge on the Fraser River, 1934.

p. 128 – The Mowats and Mutt, taken in Saskatoon, c. 1935.

p. 145 – Farley, still naked, at the Saskatchewan River, 1936.

p. 164 – Kazabazua, the station stop for the Thomson family cottage on Danford Lake, Ontario, 1937.

p. 176 – Angus and Helen in Saskatoon, c. 1935.

p. 187 – Bruce Billings, Farley's closest friend during his time in Saskatchewan, 1937.

p. 201 – Farley's owls, Wol and Weeps, 1937.

p. 215 – Bert Wilks at the Churchill, Manitoba, camp, 1936.

p. 235 – Munro Murray at the wickyup in Saskatchewan, 1936.

p. 248 – Mutt with Angus's Mutt-proofed garbage can, Saskatoon, c. 1937.

All photographs courtesy of the author

For information about permission to reproduce selections from this book, write to Permissions, Houghton Mifflin Company, 215 Park Avenue South, New York, New York 10003.

Library of Congress Cataloging-in-Publication Data

Mowat, Farley
Born naked / Farley Mowat.
p.   cm.
"A Peter Davison Book"
Previously published: Canada : Key Porter Books, 1993.
ISBN 0-395-68927-9
ISBN 0-395-73528-9 (pbk)
1. Mowat, Farley — Childhood and youth. 2. Authors, Canadian — 20th century — Biography. 3. Canada — Social life and customs.   I. Title.
PR9199.3.M68Z463   1994
808'.0092 — dc20   93-23702
[B]   CIP

Printed in the United States of America

AGM 10 9 8 7

Portions of Chapters 9 and 10 have been adapted from *The Dog Who Wouldn't Be*. Portions of Chapter 15 have been adapted from *Owls in the Family*.

The publisher acknowledges the assistance of the Canadian Council and the Government of Ontario.

For my companions of those years,
both human and otherwise;
and also, with a tip of my bonnet,
for PETER DAVISON,
who has guided my work for,
lo, these many years.

IN THE BEGINNING

Now, God be thanked, those were happy days
and we had enough sense to savour them while they lasted.

ANGUS MOWAT

On a cloudless August day in 1950 my father boarded a Trans-Canada DC-3 at Toronto's Malton Airport. The airplane was eastbound for Montreal on a course which would take it not too high above the northern shore of Lake Ontario.

Angus Mowat was always an excitable man. On this occasion he was stimulated to effervescence. Not only was this his very first "aerial voyage," it would carry him over a coast with which he had been familiar as a small-boat sailor through most of his life.

Puffing furiously on a hand-rolled cigarette, he peered raptly through the porthole. The world of his younger years unrolled below him, and he recited aloud a litany of seamarks as these hove into view and then were swept astern: "Frenchman's Bay!...Peter Rock!...Colborne!...Presqu'ile!"

Near the mouth of the Murray Canal, the plane tipped a wing as it lazily changed course. By then, Angus was squirming in his seat like a birthday child. He was over home waters now and approaching Trenton, where he had been born. When the head of the Bay of Quinte opened before him, he could no longer contain himself. He began urgently calling for the stewardess.

She came at the double and a fine, buxom lass she was too, prepared to deal with whatever dire emergency might have arisen. Angus grabbed her arm and shoved her down into the window seat with, perhaps, just a touch too much fervour. She gave him the sidelong glance of a woman who knows herself to be irresistible and remonstrated gently, "Really, sir. This is hardly the time or place."

"'No! No!" he cried with some asperity. "Look below, dammit! See that little island in the bay? That's Indian Island! And thirty years ago my son Farley was conceived in the lee of Indian Island in the sweetest little green canoe that ever was!"

Firmly removing his hand from her arm, the stewardess eased back into the narrow aisle. There she paused before replying evenly, "Congratulations, sir. That's quite an amazing feat…in any colour canoe."

ONE

The world into which I was born in 1921 had not long emerged from a terrible war which devastated much of Europe and slaughtered ten million people. Canada's contribution to that holocaust included eighty thousand soldiers, seamen, and aviators blown to bits, choked with poison gas, drowned in the grey Atlantic, or otherwise obliterated, together with two hundred thousand disabled "veterans," my father amongst them, who came home bearing visible and invisible wounds which would afflict them to the end of their days.

Apart from these unfortunates, the War to End All Wars left Canada and most Canadians in fine fettle. Trenton, a small market town in south-central Ontario, was a case in point. When the war began, Trenton had a population of about a thousand people, mostly of Scots, Irish, and English ancestry, clumped around a fine natural harbour where the mouth of the Trent River empties into the Bay of Quinte. The townsfolk mostly made their livings from quiet trade with the farms which ran northward from the bay shore until the good soil petered out in pine forests and hard-rock country.

The war changed that. Between 1914 and 1918 Trenton was transformed. A military seaplane base mushroomed a few miles to the east; the population swelled and the town feverishly engaged in war production.

The showpiece of this industrial transformation was a vast chemical plant devoted to the manufacture of high explosives. In the words of a local Captain of Commerce: "This progressive industry put Trenton square on the map of the Modern World at last."

And almost wiped it off again when, during the final months of the war, the plant caught fire and blew up. Fortunately it was distant enough so that Trenton escaped being flattened.

Angus Mowat was born in Trenton twenty-six years before the big explosion. His father, Robert McGill Mowat (called Gill), was the son of a professor of exegesis at Queen's University and a nephew of Sir Oliver Mowat, a Kingston lawyer who became premier of Ontario and one of the Fathers of Canadian Confederation. Because of these high connections, it was predicted that Gill would "go far" in the Church or at the Bar.

But Gill was a soft-centred sort of chap who loved snowshoeing, sailing, writing poetry, and contemplating

Nature, and was singularly lacking in what was then called "get up and go." However, he *was* good-looking so he married well — to the daughter of an affluent Brockville furniture manufacturer.

When I knew my grandmother Mary Mowat, née Jones, she was a tight-lipped, disappointed woman with a jaundiced view of life. Doubtless she had reason to be sour. When her husband failed to achieve even the minimum requirements for a pastor's post, and his Uncle Oliver's firm tried but failed to make something of him as a lawyer, he was brought into the Jones family business. Here he made such a hash of things that the young couple was banished to sleepy little Trenton at a discreet distance from Kingston and Brockville both.

Gill was established as proprietor of a Trenton hardware store, where he soon demonstrated that he was no more adept at commerce than at the professions or in industry. Kindly to a fault, he gave credit (and cash too) to all who asked. He also tended to be so distracted by his poetic visions as to become a threat to the general public. On one occasion, he absent-mindedly filled an order for a dozen beeswax candles with a dozen sticks of dynamite.

It was perhaps inevitable that the store would fail, but not that it would fail three times. The Mowat and Jones families bailed Gill out twice, then gave up. So he retired in his early fifties, a remittance man living on the income from his wife's inheritance.

I remember him as a slightly absent old chap with a long white moustache yellowed by the smoke from his calabash pipe. He spent a lot of time sitting in an old leather chair staring into space through rheumy blue eyes that, very occasionally, would focus on me. He always seemed surprised to find me standing by his chair, probably because his imagination was far distant, keeping company with Hiawathan braves and dusky maidens. He could have found little enough solace in his own domestic milieu, where he endured a terrible truce dictated by the iron disapproval of an embittered wife.

Apart from having to deal somehow with the stigma of his father's failures, Angus seems to have led a near-idyllic childhood. Small, lean, and wiry, he was a water rat, at home on the mashes (marshes), criks (creeks), swamps, rivers, and lakes where he became a passionate sailor of canoes, punts, dinghies, and anything else that could be made to float.

An indifferent scholar, he nevertheless passed his university entrance examinations in 1913 when he was twenty-one. He spent that summer roaming the forests of Temagami in northern Ontario as an apprentice fire ranger and in the autumn enrolled in the faculty of engineering at Queen's University in Kingston.

In 1914 the Great War began, and Angus enlisted in the army. Within a few months, he had become a participant in the blood-bath that was overwhelming France. Early in 1918, German machine-gun fire shattered his right arm and he was invalided back to Canada. When he was released from hospital, he headed west for Port Arthur, at the head of Lake Superior, in hot pursuit of the woman who would become my mother.

Helen Thomson was the youngest daughter of Henry (Hal) Thomson, who had been manager of the Molson's Bank branch in Trenton until he made some injudicious loans. After the borrowers defaulted, Hal — his career in ruins — was exiled by his employers to Port Arthur, which was the Ontario equivalent of Siberia.

During the last years before the war, Angus had been a determined suitor (one of many) of sloe-eyed, black-haired Helen. But her parents could see no future for her with the son of Gill Mowat and Helen herself was not much taken with Angus, whom she remembered as being a "very pushy little fellow." She fell in love with a young artillery lieutenant who survived the fighting only to die during the great influenza epidemic at war's end.

In the fall of 1918 Angus appeared at the Thomson home in Port Arthur. He looked exceedingly dashing in his officer's uniform, his chest covered with medals and his right arm in a sling. As a returning hero he was now much admired. Although there had been no improvement in *his* prospects, the fortunes of the Thomson family were in such decline that the elder Thomsons could hardly offer much resistance when Angus pressed his suit on Helen. Sad and lonely, in the spring of 1919 she softened and agreed to marry him.

Their wedding picture shows him in dress uniform looking very much the dashing military man. Helen looks lovely, although there is something about her which presages uncertainty. Under the faded photograph is written in an unknown hand (not hers) the piquant query: "Whither away?"

Whither indeed. Their first destination was a log cabin at romantically named Orient Bay north of Lake Superior where Angus again found work as a fire ranger. The newly-weds spent the summer "roughing it in the bush" in company with lumberjacks, black-flies, Indians, black bears, and mosquitoes.

Although, as a wounded veteran, Angus was entitled to special consideration in government employment, his useless arm proved more of a disability than the ranger service could tolerate. In the autumn of 1919, the couple travelled south to the city of Toronto, where a job had been found for Angus as a clerk in a wholesale grocery firm owned by members of his mother's family.

This was not his cup of tea. Incarcerated in a rooming house by night and a counting house by day, he worked from eight until five pushing a pen and counting things. Whether he cared to admit it or not, he was his father's son, and this was a life for which he had neither aptitude nor appetite.

After a month or two he could no longer stand it, but his mother insisted he should stay. The Firm would look after him, she said. All he had to do was be diligent and in due course he would rise. That was the point. One was supposed to make *sure* of one's future. And with his own father's fate to remember, and the knowledge that jobs for cripples were hard to come by, he felt he had to try to stick it out.

He tried. And failed. After a year of servitude, he let it be known that "the doctors" (nameless) had decreed that, for his health's sake, he must find outdoor work. Everyone seems to have believed him, and it may even have been something more than an inspired invention.

Free at last, Angus went right back to his roots. Having packed Helen off to visit her parents for a while (by this time she was pregnant with me), he set off in the autumn of 1920 for Trenton in search of a way of life which would permit him to live according to his own choosing.

Eventually he concluded that his future lay in bee keeping. This was a decision in which Helen had no part. In truth she seldom had any significant say in major family decisions then or later. She was generally phlegmatic about this, although she once ruefully told me that being married to a man who always knew precisely what was best for all of us could be trying. Indeed, Angus was always very much the captain of his own ship and, if he was not infallible, at least he thought he was.

After buying the bees and spending most of his remaining money on a second-hand Model T Ford truck (inevitably named Henry), my father went house hunting.

The post-war boom, which everyone assumed would last forever, was then in full swing. Every enterprise in Trenton from the cooperage mill to the Chinese laundry was doing about as much business as it could handle. Everything was hustle and bustle, which meant there were few houses for sale or rent, and those which were available were beyond Angus's means. In some desperation he took his troubles to Billy Fraser, the town tycoon and owner of the cooperage. Billy had at his disposal an enormous frame house of Gothic architecture, turreted and towered, which had been built by one of the lumber barons of the previous century. Now it stood untenanted, its paint peeling, its massive central tower gaping open to the weather, and its shingles falling away like scales from a scrofulous old dragon.

Billy let Angus and Helen live here by grace-and-favour, as it were, and he let Angus cart away truck loads of hardwood scrap from the cooperage to keep them warm. Nothing short of a volcano could have properly heated that old pile but, by shutting themselves into the winter kitchen and two of the servants' rooms, they at least had a roof over their heads and walls tight enough to keep the water from freezing in the buckets most winter nights.

This was Helen's first home of her own and on May 12, 1921, it became almost but not quite my natal place. As Angus described the event: "From his earliest years Farley had some disregard for the conventions. In the first instance he tried to get out when his mother was in the taxi cab on the way to the hospital, which was ten miles away in Belleville. He nearly did, too. Then, in the hospital, he'd be damned if he'd wait for the doctor, who was something of a slowpoke, so out he popped all over everything. The nurse said that the first thing she knew there he was. She said he rolled over, propped himself up on one elbow and gave her a kind of leer."

Angus was delighted to have a son, but not as delighted as he might have been.

"I much wanted a son who would become a salt-water sailor, perhaps a deep-sea captain," he complained. "Well,

Neptune puts his mark on those fated to go down to the sea in ships. They are born with a caul over their heads. This is Neptune's guarantee that they will never drown or, if they do, that they will become Mermen and enjoy a rollicking hereafter among the Mermaids."

Alas, I had not even the vestige of a caul. I came into the world just like every other landlubber does — stark naked.

We remained in Billy Fraser's rambling old ruin through my earliest years, leading a relatively uneventful life but one which was not without its moments. I was painfully slow at learning to use the pot until a night in my second year when all the ceiling plaster in my room fell down on top of me. I used the pot diligently thereafter, fearing that the entire roof would fall in if I didn't.

Our diet consisted mainly of porridge, soda biscuits, and honey. Helen relied heavily on oatmeal. Being a bank manager's daughter, she had been raised to be a lady. Four years in a convent school had taught her embroidery, how to paint with water-colours, how to declaim poetry, and how to sing in a choir. Nobody had ever taught her culinary skills, but she did learn how to cook oatmeal porridge during her summer with Angus at Orient Bay.

To celebrate our first Christmas together, she reached for the stars and determined to bake a batch of mince pies. All might have gone well had she not asked the butcher for five pounds of mincemeat "and do please cut it from a nice tender young mince."

The bewildered butcher gave her five pounds of bloody, minced (ground) beefsteak, to which she happily added all the other ingredients called for in her cookbook. The resultant pies might have been hailed as *nouveau cuisine tortière* in our time, but not in Trenton in 1921. The neighbour's dog got them and a chastened Helen returned to variations on the theme of oatmeal porridge.

Angus's bees did well at first, producing quantities of good clover and buckwheat honey, but since the market was flooded, the crop could not be sold and we had to consume much of it ourselves. I was weaned on soda biscuits soaked in milk and lavishly sweetened with honey — a dish I still find delectable.

Bees loomed large in my early years. When Angus rattled off in Henry of a summer morning to work his hives in the apple orchard of the Ketcheson farm on York Road, he would often take Helen and me along. Helen would sit under a tree and read. To keep me from crawling into trouble, she would place me in an empty super* set on the grass nearby. This was the scene of my earliest recollection.

I see, in my mind's eye, a large and strikingly marked honey bee standing on an anthill near where I sit. This bee is resolutely and briskly directing the ant traffic away from me, much as a policeman might direct members of an unruly crowd away from some important personage.

I have since been told by expert apiarists that such behaviour by a bee would be "atypical," which is a polite way of saying my memory lies. Nonsense. I *know* I was taken under the protection of the bees, and the proof is that

---

A rectangular wooden frame used in the upper portion of a hive.

I have never been stung by one, not then or ever. Wasps and hornets, yes. Bees, no. I believe I was adopted into their tribe, and ever since I have been as kindly disposed to them as they to me.

Angus had a mania for naming things, even when they already had perfectly good names. As a freshman mining engineer at Queen's he had re-christened himself Squib. In mining terminology, a squib was a small but potent charge used to detonate a major explosion and this may well have been how he viewed himself.

He was not satisfied with my name either. I had been christened Farley in memory of Helen's beloved younger brother killed in a fall from a cliff, but before I was three months old Angus had begun calling me Bunje after some character he had encountered in a novel by, I think, H. G. Wells. When my mother tearfully remonstrated with him, he airily replied that Bunje was merely my "working title" to be used until I could make up my own mind what I wanted to be called.

Because it was so forbidding, Angus initially named our decaying home The Fortress. Later he changed it to Bingen.

Bingen, as I would learn much later in life, is a pleasant little town located on the river Rhine. It is dominated by an old, square fortalice called the Mouse Tower, wherein Hatto II, erstwhile Bishop of Mainz, was punished because he taunted the poor, who were begging at his palace gates, by calling them rats and mice who would eat up all his corn.

An enormous plague of real rats and mice thereupon descended on Mainz and so terrified Bishop Hatto that he fled from the city and took refuge in the tower at Bingen. However, the rats and mice swarmed into the river, swam across it and, gnawing and tearing their way through walls and windows, cracks, and crevices, reached the inner chamber housing the terrified Hatto and ate him alive.

The myriads of rats, mice, bats, and other creatures which made free with *our* rambling ruin were friendly fellow inhabitants. Their presence may even have helped give rise to the sense of fraternity with other animals which has so powerfully influenced my life. Certainly they were no cause for apprehension.

Bears were a different matter.

One night during our final winter at Bingen, my bedroom was visited by an enormous bear. I woke to find him standing upright by the window. He was wearing a checked tweed cap with matching visor and staring about him as if in surprise and even confusion. He did not seem inimical but I was nevertheless too startled to move a muscle. It was not until he began to shuffle towards the door that I let out a yell which, in my father's words, "brought your mother and me up all standing!" They were in my room within seconds but arrived too late to see my visitor, who had hurriedly decamped.

Although my parents assured me it had all been a dream, that bear was as real as anything could be. I can still see him in awesome detail: about seven feet tall, brown-furred, long-clawed, and, except for the rakish cap he wore, truly *bear*. He startled the hell out of me but I couldn't have been too frightened because I was content to remain alone in that room not only for the rest of the night but for our remaining time at Bingen.

I even harboured hopes he might return. This time I would be expecting him and, who knows, we might become friends. But he never came back, and I think I know why. My parents were partially correct. It was a dream all right, but *it was the bear's dream*. And I think we scared the bejesus out of *him*.

TWO

During the summer of 1923 the apiary, which had always been a losing proposition (we kept bees…not vice versa), was smitten by a pestilence called "foul brood." The bees perished in their tens of thousands, leaving us without even enough honey to spread on the soda biscuits we could no longer afford.

The winter following the foul brood disaster was a tough one. Angus sought work and Don Fraser, brother of the ubiquitous Billy, tried to employ him as an insurance salesman.

"It wasn't any good," Angus remembered. "I was too shy, you see. When I saw a likely prospect coming, I'd cross to the other side of the street. But once old Tommy Potts tracked me down. He was eighty-seven, blind in one eye and couldn't see out of the other, and had halitosis that could knock a horse off its feet at fifty yards. He said he was dying and *needed* life insurance. I sold him fifty thousand or so dollars' worth on credit but Don wouldn't honour the sale."

To make things worse, the chimney in Bingen's cavernous kitchen caught fire and collapsed. We escaped unscathed but had to seek refuge in another decayed structure, which Angus christened the Swamp House because it stank pervasively of rotten wood.

My mother, most long-suffering of women, was able to endure this except when visitors were expected. Then she would burn quantities of brown paper in the kitchen stove with the chimney damper tightly shut, thereby filling the house with acrid smoke. She admitted that this was exchanging one stink for another but hoped her half-asphyxiated visitors would at least be unaware of the underlying stench of mould which, to her mind, was synonymous with the "stench of poverty."

At this juncture some of our little family's well-wishers came to the rescue. The librarian of the Trenton Public Library, a crotchety spinster who had run her little fiefdom with an autocratic hand for thirty years maintaining her position by threatening to resign if anyone questioned her rule, chose to make this threat once too often. The chairman of the board took her at her word and offered the job to Angus, at the munificent salary of five hundred dollars a year.

So Angus began the career which was to engross him for the rest of his working days. And we three began to eat regularly. Leaving the Swamp House to the mould and mildew, we took up residence in two upstairs rooms rented from Mrs. White, a railroad worker's widow whose daughter had been one of my father's girls in high-school days. This choice did not please my mother. Years later she was to tell me, "He was as charming to that little sniff of a daughter as if she was a princess; and the way she looked at him was enough to make one ill." Even then I think my mother had begun to suspect she had married a rover.

My father's interest in women was surpassed only by his passion for boats, a passion he was determined his son must share. When I was a year old, he began taking Helen and me on weekend excursions aboard one or other of the several local boats owned and sailed by friends.

In July of 1923, we embarked on our first family cruise — in an ancient sailing canoe borrowed from a retired banker. It must have been built by one of Noah's sons, and hadn't been near the water since. Its sail was so thin and sere you could see through it. Nevertheless Angus dumped a little tent, some duffle and food aboard, and we set sail.

Despite the canoe's fragility, Angus drove it hard for two days towards the eastern reaches of the bay. We sailed by day and camped on the low, mosquito-haunted shores at night. By the third day, we were half-way along the coast of Big Island when a nasty black squall with thunder and driving rain hit us over the stern, and Angus made for land.

None too soon either, according to Helen. "You were like a little wet rat," she told me many years later, "too cold even to cry. I was crying, with fury. Angus knew that *I* never liked sailing except for short runs on sunny days with *very* little wind. But you know what your father was like when he was determined to do something dramatic."

I did not know at the time, of course, but I learned that when it came to engaging in sheer, pig-headed histrionics, Angus had few rivals. This was a characteristic which caused my mother much distress, although it did not bother me during my childhood years.

The following day it continued to rain heavily, with great gusts of wind which finally dissolved the old sail into flapping fragments. There being nothing else for it, Angus reluctantly (and with difficulty) managed to paddle us into Picton harbour. Here the voyage ended in a mutiny. My mother simply refused to go any farther in "that leaky old thing, and Bunje just a two-year-old."

Angus was then left with no alternative but to take a train to Trenton to pick up Henry, and come and get us and the canoe. The loss of face he suffered or thought he suffered in having to end the voyage so ignominiously was something for which I think he never quite forgave my mother.

That autumn we had to move, again. Very late one rainy November evening Angus came home from duck hunting. Instead of rousing someone to let him in, he climbed up on the roof of the porch and set about removing a storm window from his and Helen's bedroom. Not surprisingly, he lost his footing on the wet shingles and fell to the ground still clutching the storm window. The resultant crash wakened the whole household and most of the neighbours. As my mother remembered it, "He was a little under the influence, you know. So he wasn't hurt. But then Mrs. White's shameless daughter rushed out keening like a Valkyrie and threw herself on him as he lay on the grass, and next morning Mrs. White gave us our marching orders."

Our next home was a two-bedroom apartment above a clothing store on the main street. This was not a select residential neighbourhood but the rents were low and it was adjacent to the harbour, which was convenient since Angus had decided to buy a boat of his own.

He bought a seventeen-foot Akroyd sailing dinghy — a beautiful, varnished centre-boarder that sailed like a witch, and cost him a quarter of his first year's salary. He christened her *Little Brown Jug** and she became the apple of his eye.

Our first cruise in *LBJ*, as she was familiarly known, took place in the summer of 1924. It should have been a pleasant saunter through the sheltered waters of the Bay of Quinte to Kingston and have taken no more than four or five days. It took ten. Angus's log chronicled the voyage with the laconic insouciance of a master mariner taking his square rigger around Cape Horn. Helen, however, recalled it with the kind of shuddering horror which might have afflicted a French gentlewoman being conveyed to the guillotine in a tumbril. I recall it as through a glass dimly, and wetly.

Inspired by a popular drinking song, "Little Brown Jug, oh I love thee."

It was an exceptionally cold summer. Thunderstorms occurred almost daily, accompanied by vicious squalls that churned the shallow waters of the bay into yellow foam. On one occasion the wind blew with gale force for thirty-six hours. High winds and high waters produced floating debris ranging from tree trunks to a dead pig. One day a piece of flotsam was driven into the centre-board housing, jamming the board in the "up" position. This was no problem as long as we were running free, but when a howling head wind suddenly burst upon us, *Little Brown Jug* could get no hold on the water and was blown into a vast cat-tail swamp behind Foresters Island. Here for an entire night Helen endured a local version of the travails of Katharine Hepburn aboard *The African Queen*. Meanwhile, Helen's Humphrey Bogart flailed about in the swamp muck under the hull trying to free the centre-board. Had this in fact been Africa, Angus would have been a goner. The crocodiles would have got him. And if they had, I suspect my mother might not have mourned overmuch.

LBJ was a racing machine built for day sailing. She had no cabin and offered no more shelter to man, woman, child, food, or bedding than could be found in a cubbyhole under her tiny foredeck. There was barely sufficient space for me in this cramped little cave. And it seemed never to stop raining or, if it did, the wind blew so hard that spray soaked everything anyway.

I recall sitting on the floorboards with bilge water sloshing over my bare legs and around my naked bum. More water dribbled down through a seam in the foredeck a few inches above my head, and still more came pelting in through the entrance to my cave whenever a rain squall drove against the sail. Being themselves soaking wet most of the time, there was little enough my parents could do for me.

Helen put a thick woollen sweater on my top but there was no use clothing the rest of me, which was sitting in the soup. There were some crayfish crawling around and I amused myself playing with them.

"When I peeped in at you," Helen remembered, "I thought of Water Babies and wondered if you might sprout gills."

I appreciated those crayfish, and also some pollywogs, but I doubt that I was particularly happy slithering helplessly about on the slippery floorboards whenever *Little Brown Jug* came about, or being jounced mercilessly up and down as she butted her way into what must have been quite formidable waves.

The experiences I endured during this primal cruise were indelibly etched into my subconscious. Alas, my poor father! In attempting (in his spartan way) to inculcate a love of the sea in his son, he but succeeded in instilling in me a deeply rooted distrust of the sailing life, as anyone who has read *The Boat Who Wouldn't Float* will know.

The immediate result of that ordeal was a protracted trial of wills between my parents. Although occasionally and reluctantly inveigled into going day sailing, Helen resolutely refused to go cruising again unless in a much larger boat, one with a comfortable and waterproof cabin, and an engine that could hurry the vessel into safe harbour in times of storm and peril.

She was not being selfish. Selfishness was no part of her nature; she was worried about the perceived threat to *my* survival. Although prepared to be almost infinitely malleable on her own account, she could become an intractable obstacle if she felt my well-being was in jeopardy. "Balky as a bloody mule!" was how my disgruntled father put it.

Being used to getting his own way, Angus held out for an entire year. Then, near summer's end of 1925, he capitulated. He sold *LBJ* and bought an antiquated twenty-six-foot, Lake Ontario fish boat. She was propelled by a 10-horsepower, single-cylinder gasoline engine and did not even have a sail. She was, in fact, that ultimate anathema of all true sailor men — a *"stink-pot!"* *Angus Mowat had bought a power boat!* Neptune surely shuddered. This was not mere capitulation; it was abject surrender. Or so it seemed.

"Your father was such a cunning man," Helen remembered sadly. "The artful dodger!" He *did* build a cabin on *Stout Fella*, (so-called because she was) and it was quite comfortable. But he either couldn't or *wouldn't* make the engine run properly. It was always stopping at the most awkward moments, leaving us drifting about for hours until some kind soul gave us a tow. Angus didn't say much about these contretemps, except to swear at the engine, but sometimes he would mutter loudly enough for us to hear, "If only this was a sailboat, we could get home on our own."

One summer day in my fifth year, *Stout Fella* was belching her noisy way eastward down the bay towards the combined causeway and bridge which connects the almost-island of Prince Edward County to the town of Belleville on the mainland.

The movable central span was already swinging to allow a west-bound tug towing a string of coal barges to pass through. Angus concluded (or so he told us) that there would be ample time for us to slip through ahead of the tug, so he headed *Stout Fella* for the gap. Just as we entered it, the engine gave a terrible backfire and quit.

When the engine failed, I was standing near the bow feeling superior to the passengers in a long line of motor vehicles and horse-drawn wagons backed up on both sides of the swinging span. *Stout Fella* lost way and did a slow pirouette until she lodged sideways across the gap — her bow aground on some buffer logs edging one bridge pier and her stern jammed against the casing of the opposite pier.

The tug (the *M. Sicken* out of Trenton) sheered off from the gap, her hoarse whistle giving full vent to her skipper's outrage. Car drivers began to blow their horns. The bridge master, a retired lake captain with a flowing white beard, shot out of his little cubicle and hung over the railing ten feet above our heads.

"Gol durn you, Mowat, you done that a-purpose!" he bellowed. "Now you git that old strawberry crate out of there or by the livin' Jesus I'll have the Belleville garbage truck come and git ye!"

His irritation was warranted. This was a hot Saturday morning and the causeway was full of produce-laden farmers' trucks bound for the Belleville market. The raucous blare and tootle of their horns filled the air. Horses neighed their distress. Red-faced men and women stomped from the vehicles towards the gap, angrily waving their arms at us.

In the face of all this hostility, I retreated uncertainly to the cockpit. Humiliated beyond endurance, Helen burst into tears and fled into the little cabin, slamming the companionway hatch behind her. The comments from above became ever more derisive.

"Why'n't you just pull the plug and let that bathtub sink?" someone shouted.

"Ain't no bathtub! Looks more like grandad's privy what went adrift in the Big Storm last fall," jeered another.

One particularly irate farmer, who had been to school with Angus, shouted, "Thought you was supposed to be a sailor, Mowat! What in hell are you doin' driving a goddamn *tractor*?"

The completely uncharacteristic way Angus endured this barrage leads me to believe the bridge master was correct. Nary a sharp rejoinder crossed my father's lips as, calmly and unhurriedly, he took the boat-hook and worked the vessel free. Then, almost gaily, he leapt ashore on the rocky rip-rap with a line in his hand and hauled *Stout Fella* out of the gap, clearing the way for the *M. Sicken*, whose safety valve seemed about to pop. Only then did my father respond to his tormentors.

"That's right, Johnny. I *should* have stuck to sail. Won't make *that* mistake again." And he smiled sunnily up at the crowd as its constituent parts began to head back to their vehicles.

He was still smiling as he turned to me and said, "All right, Bunje-boy. You can tell your mother it's safe to come on deck again."

———————————

That winter Angus rigged *Stout Fella* as a ketch and thereafter we sailed her almost every spring, summer, and autumn weekend, and for as much intervening holiday time as Angus could squeeze out of the library board. Since the board included three other dedicated sailors, it was generous in this regard. It was less so with money, of which in truth it had only a pittance to dispense. But we happily made do on an income which by current standards would be well below the poverty line.

We led a good life, no small part of which was lived afloat at little cost. Although we did use the "bullgine" occasionally to get us out of difficulties (it never again failed us, be it noted), for the most part the wind provided free fuel. Food was to be had for the taking (fish from the bay), or the asking (vegetables, milk, cream, butter, and eggs from the many farms along the shores). Farm wives would often give my mother fresh-made bread and pies, jars of preserves and pickles, bottles of maple syrup, a chicken or a ham, or a cut of fresh meat if an animal had been slaughtered recently.

These people would have indignantly refused money in recompense but Angus was able to reciprocate in his own way. Although most of the county farmers were passionately fond of reading, books were always in short supply, so Angus began surreptitiously lending them volumes from the

Trenton Library. *Stout Fella* became a kind of forerunner to the Travelling Library trucks which now serve rural regions.

There were scores of sheltered coves and anchorages around the bay and *Stout Fella* came to know them all. I came to know their people: dairy men, apple growers, commercial fishermen, family farmers, poachers, pot hunters, village merchants, even one or two moonshiners. For the most part they were United Empire Loyalist stock — people of Dutch and English ancestry who had fled north from the Thirteen Colonies as refugees from the American Revolution. They were people of conviction, of enduring loyalties, and of great generosity.

One of our favourite haunts was Prinyer's Cove, a tree-shaded slit in the Prince Edward County shore where we would lie lazily at anchor through days of summer content. Usually we had this hidden place to ourselves. The glittering, snarling hordes of mass-produced floating automobiles which now roil the placid waters of the bay were, as yet, unknown.

It was my job to row ashore early each morning through a dawn mist and pad barefoot up the dusty track to a nearby farm to collect a can of milk still warm from the cow. Generally I would also scoff a preliminary breakfast with the farmer's wife and her three daughters. Back aboard *Stout Fella*, I would have my second breakfast: oatmeal porridge slathered in creamy milk and drenched with maple syrup or mounded with brown sugar. Everyone knew sailors had to have hearty meals, and diabetes was not something we worried about.

At some of our "ports of call," we would moor alongside a wharf or jetty, as at Rednersville, a village consisting of three houses, one general store, and a rambling canning factory thrusting a skinny iron chimney high into the air. At harvest time, which began when the first field peas were ripe, this chimney would spout a great plume of sulphurous coal smoke, the signal for scores of wagons and old trucks laden with canning crops to converge on the plant.

Many farm kids would come along for the ride and they made festival about the dock, from which one could dive to a cool, dark depth of ten feet or more. They were frankly envious of me and "my" boat, and once a boy of about my own age offered an ancient penny (which he claimed was pirate gold) if I would let him stow away aboard *Stout Fella*.

Although it was through our voyages that I really came to know the local world of waters, my baptism as a water baby had taken place shortly after my birth.

The summer of 1921 had been exceedingly hot and dry. So Angus's old friend, Norman Kidd, who had a cottage on the south shore of the bay not far from the mouth of the Murray Canal, invited us to come and stay with them in what was to be the first of many visits spanning several years.

The Kidd cottage was a cavernous, roughly built, almost windowless, frame structure originally constructed by

Norman's father as a duck-hunting camp. Its walls were plastered with enormous, highly coloured posters depicting ducks, geese, bears, moose, et cetera, being slaughtered by intrepid sportsmen. Even as a very young child I felt the chill of death in this gloomy space. Fortunately, we three had to enter it only to have our meals. We slept in a screened-in, canvas-roofed cabin where we benefited from whatever breeze might waft in off the bay.

The wayward little cluster of shacks and cabins composing "Kidds' Cottage" included a fine ice-house. In the days before rural electrification, almost every family around the bay had one of these to supply the kitchen ice-box. Every winter tons of ice blocks were hand-sawed out of the frozen bay, hauled ashore in horse-drawn sleighs, and stored between layers of sawdust in garage-sized structures whose walls and ceilings were also thickly insulated with sawdust.

Ice-houses were a special summer domain of children. During the blistering heat of August days, we youngsters would spend hours cooling our bottoms and our bare feet in the wet sawdust while the crickets whirred outside. The Kidds' ice-house was a chill, dark sanctuary where the imagination was free to create worlds of one's own. Sometimes the place would become a polar bear's den and we the bear cubs. Sometimes it was an Eskimo igloo. Once it came perilously close to becoming a tomb.

The thick, insulated door latched on the outside. One day when four or five of us children were inside, someone slammed the door and we were locked in. The only illumination was a wan ray of light from a small ventilation shaft in the roof, which did little to lighten the clammy, chilly gloom.

For a time we were excited by our predicament, seeing it as a novel adventure. But when nobody let us out and the cold began seeping through our thin summer clothing, we grew frightened.

At seven years of age, Jack Kidd was the eldest. He drew us together in a shivering cluster just under the ventilation shaft and led us in a cry for help that soon degenerated into tearful howls. Although we yelled and cried ourselves hoarse, no one came. Nobody heard us because of the thick insulation in the ice-house walls.

We huddled tightly together in the damp sawdust and the cold bit deeper into our bodies. For the first time in my life I felt terror. One of the little girls had buried her face in my neck and was sobbing bitterly. I was probably weeping too. I know that some of the nightmares which followed upon this incident (and did not cease until I was in my teens) made me cry so despairingly that I would wake with my face drenched in tears.

By the greatest of good fortune, someone from a neighbouring cottage came by to get some ice, and released us. It was none too soon. One little boy had to be attended to by Dr. Farncombe for what would doubtless now be called hypothermia. I suffered no lasting ill effects but to this day grow uneasy in cold, dark places.

I don't think I went swimming during my first visit to the Kidds' but a photograph taken in the summer of 1922 shows me sitting in the landwash up to my naked navel. By the time I was four, I could dog-paddle well enough to swim with the other children, unsupervised by any adult. We spent hours every day mucking about in the warm waters, chasing frogs, water snakes and crayfish. We did not mind when, in late August, the bay became awash with a floating green slush called Dog Days — an explosive "bloom" of green algae. I remember the joy of diving under and surfacing through it to emerge shrouded and streaked with tendrils of green slime like a water demon.

Norman Kidd provided a little skiff for us. By the age of four I could row it handily. At five I rowed, all on my own, a half-mile offshore to Indian Island* — while Angus proudly cheered me on and Helen wrung her hands and bravely refrained from calling me back.

Indian Island had earlier been known as Massacre Island in memory of a party of Hurons reputedly killed on it in the eighteenth century by an Iroquois raiding party. It was a place of delicious possibilities. I found bones (possibly human) washed out along its shores, and crayfish could be caught by flipping over the slippery flat stones in the surrounding shallows. Freshwater clams lurked in waters shoal enough for us to reach them and sometimes contained tiny "seed" pearls.

Half- or altogether naked, I lived the life of a water baby during those halcyon days. In my father's time, people who messed about by and in the waters of the bay proudly

---

*The very same Indian Island under whose lee, according to Angus, I had been conceived. Was my conception calling to me?

called themselves Bay of Quinte Bullfrogs. Regardless of where life was to later lead me, I believe I am at least entitled to style myself a Bay of Quinte Tadpole.

THREE

Trenton had been a major port during the days of the timber trade and continued to prosper into the early twentieth century, shipping barley across Lake Ontario in big sailing vessels. However, by the 1920s the schooners were all gone, and only a few colliers still called, together with the occasional steam packet carrying freight and passengers between Toronto and Kingston.

Sometimes when one of the packets puffed into harbour, Angus would take me aboard her. I was allowed to clamber around freely, both above and below decks, while he gabbed with officers and crew, some of whom he had known since his own childhood. Although I never felt any urge to become a sailor, my visits to the old vessels gave me an admiration for working ships and their people which still endures.

On one occasion Angus and I shipped aboard the tug *M. Sicken* as guests of Captain Ben Bowen on a run from Trenton to Belleville. As we approached the Belleville bridge, *I* was the one who pulled the whistle lanyard to summon the bridge master to his appointed task. Small and shabby, for she was more than half a century old, the *M. Sicken* spewed black coal smoke like a volcano, coating herself and everything around her with gritty dust and ashes. But to me she was Leviathan.

The *M. Sicken* may have been responsible for my first appearance in print. Having become enamoured of the Pooh books, I wrote a letter to Christopher Robin, enclosing a picture of myself dressed in a sailor's suit standing on the *M. Sicken*'s bridge. To everyone's astonishment but mine, Christopher Robin replied. He had been much impressed by the *M. Sicken*, which he seems to have thought was my personal yacht. Our two letters were reprinted in the *Trenton Courier*, thereby giving me an early taste of literary notoriety.

The west side of the harbour was dominated by the massive bulk of the cold storage plant whose four-square, windowless limestone walls towered like those of a mediaeval fortress. Built in the mid-nineteenth century to store apples and other perishable farm products awaiting onward shipment by water, the vast stone vault was by now largely abandoned and had acquired a forbidding air of decrepitude which irresistibly attracted youngsters. Exploring its dank recesses was as good as exploring the dungeons of a haunted castle.

It also attracted the town drunks and ne'er-do-wells who used it as their club. Lolling at ease on piles of old sawdust, they consumed (when they couldn't get anything less lethal) a solidified form of alcohol called Canned Heat. We children were allowed to consort with these ragamuffins, and I especially remember a lanky redhead called Bunny-Boy who had once worked in a circus and could juggle half a dozen tins of Canned Heat at a time. These men also carved toy boats for us and showed us how to catch catfish on trotlines. If they posed any sort of threat, sexual or otherwise, we never knew it.

We fished whenever and wherever, but nowhere more assiduously than from the harbour wharves. Our catch was mainly perch, rock bass, and sunfish — "panfish" — which Angus encouraged me to bring home as my contribution to the family's larder. However, Helen, who had to clean and scale the bony little creatures, *discouraged* me. I think this was my first experience of being caught in the middle of the battle of the sexes. I compromised. I would bring home *half* my catch and give the other half to the neighbourhood cats.

Many fishermen still made a livelihood on the bay working single-handed from small open boats. Their rickety wharves and spindle-shaped net-drying racks seemed to be everywhere along the shores. A crony of my father's by the name of Milt fished out of Onderdonk's Cove and occasionally took me with him when he hauled his nets and lines or lifted his fish traps. Like most of his ilk he was a "general purpose" fisherman, catching lake trout, white fish, black bass, pickerel, pike, eel, smelt, and something called a sheepshead. This was a large and coarse-scaled fish chiefly remarkable for its voice. While sitting silently with Milt in his boat, I would sometimes hear loud grunting sounds from the depths below. "That's old man sheepshead talkin' to his woman," Milt would say.

Of course I did not spend all my childhood hours on or near the water. In winter we would go trundling noisily around the countryside in Henry, visiting the farms to which Angus brought the benison of books.

We were rewarded with gargantuan farm meals and with sleigh and cutter rides across frozen marshes and ice-bound ponds into the deep recesses of cedar swamps and hardwood forests. Here I saw deer, snowshoe rabbits, ruffed grouse, and foxes. On one occasion we encountered a lynx which paused in the deep snow fifty feet away to stare through slit, green eyes at half a dozen human beings staring back at it over the steaming back of a big Clydesdale. Experiences like this fuelled the fascination I was beginning to feel for animals.

A favourite visitation of mine was to Charlie Haultain's fox "ranch," which consisted of half an acre of waste land on the edge of a swamp. It was surrounded by an eight-foot fence roughly made of sawmill slabs set on end. This stockade enclosed a dozen fox pens together with the shanty where Charlie lived. In one corner of the enclosure stood a tower of peeled poles about thirty feet high supporting a tiny cabin reached by a rickety ladder from which I could observe the foxes unseen by them. It was like having a window into their secret lives and I was happy to spend hours watching them eating, at play, and making love.

The establishment may have been a "ranch" in Charlie's eyes — the imagery of the Far West loomed large to young men in those days — but for me it was more like one of the fur trader's forts portrayed in a picture book about the old-time voyageurs which I had unearthed in the library. Charlie fit the picture. He was swarthy, swift of movement, and familiar with everything that lived in the fields, woods, and waters. He would have been perfectly at ease clothed in buckskin with feathers in his hair.

In fact, he was generally clothed in the powerful aroma of dog-fox, which is similar to skunk. It permeated his cabin, and his garments, which he did not often change. I thought the aroma rather marvellous (it smelled to me then, and still does, like the essence of wilderness) but most Trentonians preferred to keep Charlie at a discreet distance. He and Angus also took me snowshoeing in the woods, and Charlie declared that, given time, he would make a real woodsman out of me. Unhappily, when I was six he lost interest in fox ranching and went off prospecting for gold in the far north.

Another of my father's cronies was Vic Bongard, who was so dead keen about sailing that he sailed summer and winter. Vic owned an iceboat — a cross-shaped hickory frame about twenty feet long by twelve wide, fitted with large skate blades at each extremity. Driven by a disproportionately large sail, this contraption could skim over the surface of the frozen bay at forty miles an hour.

One brisk Sunday afternoon in February, Vic and

Angus somehow persuaded Helen to join them on a little run across the bay. I went too, and we had a splendid sail to Rednersville where we went ashore to visit a farm family. The farm children and I drank sweet cider while our elders drank the real stuff, to such effect that they were persuaded to stay for a goose dinner which lasted until nearly midnight.

By then the weather had changed and snow was beginning to fall. When we set out for home we could no longer see the lights of Trenton some seven miles away, but a glimmer of moonlight was still filtering through the clouds so off we sailed, my mother and I well bundled up in an ancient buffalo robe. The snow began to fall more and more thickly and the darkness deepened, but the breeze held steady and on we skimmed, the blades swishing with the sound of giant scythes.

Then we stopped. To be more accurate, the iceboat stopped. We four continued on like hockey pucks vigorously propelled over new ice. Clinging to each other and half smothered by the robe, Helen and I must have slid a hundred yards before we managed to gain our feet. I was wildly exhilarated. Helen was as mad as I had ever seen her.

"Oh, you *fools!*" she shrieked at Vic and Angus as they emerged through the thickening murk. "You absolute *idiots* could have drowned us all!" She strained for the right epithet. "You . . . you *MEN!*"

It was no use pointing out to her, as Angus tried to do, that there was a yard of solid ice beneath us. That only made her madder. She stamped her foot. "Oh, what's the *difference!* If Farley and I had hit a tree *we'd have been as good as drowned!*"

The iceboat had, in fact, hit a ridge of snow scraped up by somebody cutting ice blocks. Angus and Vic said nothing about it at the time but it occurs to me now that if we had sailed into the channel the ice cutters had made, my mother's awful prophecy might well have been fulfilled.

The Bonter farm, not far from the village of Consecon, was a favourite place to visit in summertime. Elmer Bonter kept dairy cattle and his wife made Devonshire cream. Served with fresh strawberries or raspberries, and topped with dollops of buckwheat honey, this was a dish the memory of which can still make me salivate like a Pavlovian dog.

I have another memory of the Bonter farm. One day I allowed myself to be dared by the Bonter kids into trying to walk across the farm manure pile.

Inevitably I broke through the crust and sank into it up to my armpits. When I was pulled out by an irate Elmer, I had lost not only my shoes but my trousers too. Even prolonged immersion in the nearby waters of Lake Ontario, laving myself with home-made brown soap, could not wash away my feelings of humiliation.

My other companions of those early days are now mostly lost in the mists but I recall a couple with some clarity. One was the son of a local doctor. Now spending his days on a Florida beach like so many Canadian jellyfish, Doug Reid, who witnessed my humiliation at the Bonters' farm, today complains it was my predilection for running around naked which led him to choose a nudist colony as his retirement residence. He says I had a bad influence on him.

The other friend was the son of a poacher who lived a bachelor existence in an old boathouse that sometimes went adrift during spring high water. I have forgotten the boy's name but Helen used to refer to him tellingly as the Marsh Boy. He was a child of few words who generally avoided human company. But we were wharf-rats together when I was five, playing on and under the decaying old steamer docks, in the hulks of abandoned barges, or poling a punt around in the swamps. He led me even farther into the world of the other animals by showing me a bittern's nest; muskrat houses; the nesting haunts of snapping turtles; giant carp that lay half awash like crocodiles amongst the reeds; bullfrogs; and the thick, black water snakes who lived under his floating home.

I was enthralled by him and the world he lived in and took him home to lunch one day so I could show him off to my parents.

He was ill at ease as we climbed the stairs leading to our apartment over the clothing store, and he baulked entirely when he got to our open door. I think he would have fled had not my mother come forward, beaming in welcome, to take his arm and urge him in. She offered him a steaming plate of macaroni and cheese, which he turned down with something very like a snarl. Staring suspiciously at my father through the ragged tangle of black hair which hung over his brow, he backed away from the table, pulled a knotted handkerchief out of his pocket, untied it, and spread a clutch of hard-boiled eggs on the kitchen floor.

These were not the products of your ordinary domestic hen. One was a heron's egg; a couple had probably belonged to a hell-diver; and one very large, olive-brown one might have been laid by a gull. While we watched, fascinated, the Marsh Boy systematically cracked each egg against his forehead, peeled off the shell, and gulped down the contents. Having had his lunch, he departed with no further social parley.

Although I pressed him, he would not again risk his liberty to the confines of our apartment. That may have been just as well since one of his favourite snacks was frogs' legs. Eaten raw. I doubt if my mother could have handled that.

Once the sailing season began the Marsh Boy faded from my life. I didn't see him again until the autumn when we both began attending Grade 1 at Lord Dufferin School. By then I had almost forgotten our close kinship of the spring, but he had not. One day at recess he gave me a painted turtle.

This was a phlegmatic but indomitable creature the size of a small dinner plate. He was allowed to range around the apartment at will, demonstrating his bulldozer power by crawling under heavy objects such as a coal bucket and sliding them along. To the consternation of a friend of my mother's, he once slid a basket of washing right across the kitchen floor without ever revealing his presence under it.

Hercules, as I named him, delighted in joining me when I had a bath. He liked to be on top of things and would stand peering myopically over the edge of our kitchen/dining table for hours, seemingly contemplating the depths of an imaginary pool into which he never quite dared plunge.

One day just before Christmas he was ambling around the table top where I was busy drawing pictures. Suddenly he gave a cry — a sort of sonorous honk which startled both Helen and me. As we stared at him, "he" laid an egg, shuffled to the table's edge, and plunged to the kitchen floor. The fall caused no apparent injury but thereafter Herc became strangely withdrawn, even for a turtle. She took to spending most of her time under my bed. She would have nothing to do with the egg, a leathery, lozenge-shaped object, so Angus buried it in a box of moist sand and placed it in the warming closet of our big coal stove. It never hatched, and in the spring we gave the unfulfilled Hercules her liberty in the swamps of her birth.

As for the Marsh Boy, after the first week or so he failed to return to school. I never saw him again nor, I regret to say, did I make any effort to seek him out. What unconscious cruelty we practise as children!

Although the eccentricities of some of my pals did not perturb Helen, whose tolerance seemed almost limitless, she did worry that I was, in the words of her younger brother Arthur (who was only six years my senior), such a "skinny little wart." Certainly my appearance was not prepossessing. At the age of five I weighed only about thirty pounds, had arms and legs that made match sticks seem robust, and balanced a pumpkin head on a neck as graceful and nearly as thin as that of a stork. Kindly people said I looked delicate or frail. Less kindly ones described me as peaked or puny. Heartless ones stigmatized me as a sickly runt — though sickly I was not.

None of this bothered me in those early days but it did distress my mother. She pestered old Dr. Farncombe for tonics, potions, and procedures which would turn me into a young porker. By the time I was entering my sixth year, he grew tired of her importunities and made an appointment for me to be examined in Toronto by Dr. Alan Brown, Canada's foremost pediatrician.

We took the train up from Trenton on a Friday morning and then a streetcar to his office, which was all very hoity-toity, with many nurses in starched uniforms running about. There must have been a dozen mothers with their children, all waiting to see the Great Man, and we joined them. After a very long time, a nurse came and took me into the inner sanctum where I was stripped to the buff and poked and pinched until I was ready to make a bolt for it. I was gone such a long time that Helen had almost given up hope for me. At last Dr. Brown himself brought me back into the waiting room.

"Who's Mrs. Mowat?" he cried.

My mother timidly responded, whereupon he shouted in a voice I'm sure could have been heard all over the building:

"Well, Madam, what do you mean wasting my time? This boy is as healthy as an ox. As for his size — if you'd wanted a prize fighter for a son you should have married one. Good day to you!"

FOUR

Angus was possessed of furious energy, with ambition to match, neither of which had ever been fully unleashed until he was hired to rejuvenate Trenton's library. Thereafter he dedicated himself to becoming the ideal librarian. He sent for and devoured everything available about what would later become known as library science. He swept through the library stacks like a cyclone, ruthlessly discarding the accumulated literary dead wood of generations. Bit by bit, for money was scarce, he then refilled the shelves with books people would take home and read. Circulation doubled, then tripled. And word got around.

In the summer of 1928 he was invited to take charge of the Corby Public Library in neighbouring Belleville, and he accepted. Belleville was the county seat and considerably larger and richer than Trenton. So was its library, which even rated a part-time assistant. Apart from the greater prestige which the move brought him, Angus's salary nearly tripled. We now had our feet firmly planted on the ladder to middle-class success.

The library building, which had originally housed the Merchants Bank, was an austere, three-storeyed limestone monument standing on the terraced slope of a hill overlooking the Moira River. A high retaining wall behind the building was full of nooks and crannies which delighted me by providing a lodging house for innumerable birds, mice, squirrels, and snakes.

We lived in a high-ceilinged, wide-windowed apartment occupying most of the second floor, which had once housed the bank manager and his family. These were by far the most spacious quarters we had yet known and our scanty collection of hand-me-down furniture did little to dispel the illusion of camping in an abandoned automobile showroom. However, the vast interior spaces were great for games of tag or tricycle riding on rainy days.

Directly across the street from the library stood the ornate brick mansion of Dr. Sobie. There were two children in the Sobie family — Jean, who was a year or two older than I, and Geordie, a dour, dark-haired lad of my own age. He and I soon became pals.

Behind the Sobie home was a big carriage house. Its lower storey had been converted into a garage but the one-time hay loft remained as a dark and echoing vault full of bats and barn swallows. Geordie and I made this our own private world. One corner of it was cluttered by scores of gallon jugs in brown, green, and dark red glass. These had once contained cough syrups and tonics, bought in bulk and dispensed from Dr. Sobie's office.* Most still had their corks in place and it did not take us long to discover that these usually contained a few spoonsful of syrup, either cherry- or peppermint-flavoured. What we did not know was that some of these potions were laden with codeine or other narcotics. On one occasion, after polishing off the residues in a number of jugs, we fell asleep in the loft and were posted as missing for several hours. Our distracted parents never suspected we'd been on a trip. Sometime later Geordie and I got so high on the alcohol which was a major ingredient of most medical elixirs that we were emboldened to squeeze through a ventilator onto the carriage-house roof, from which precarious vantage point we had to be rescued by the town fire department. By then we had begun to suspect that there were strange genii in some of the bottles.

Jean Sobie gave me one of the worst moments of my early life. Geordie and I were playing ping-pong with Jean and some of her girlfriends on the Sobie porch one hot summer evening in 1929. Suddenly the girls clustered together, pointed their fingers at me, and began tittering insanely. I was wearing very short shorts that day and no underwear. As I leapt about at our end of the table, Jean had caught a glimpse of my penis.

---

*Prohibition had been introduced into Ontario during the war and reigned until 1927. It did not reign supreme. "Medicinal alcohol," which could be obtained with a doctor's prescription, eased the thirst of many. Others drank the alcohol-based nostrums dispensed by doctors.

"*We* saw your *dinkie!*" crowed this obnoxious little female.

I hotly denied it and tried to persuade the girls that what they had seen was the handle of my ping-pong paddle. To no avail. I fled home and my embarrassment was so acute that I stayed away from the Sobie house for days. I don't think it was outraged modesty that distressed me so. I suspect it was the primordial fear which haunts most men — even when they are little boys — that their organ is smaller than it ought to be.

Geordie and I often prowled the shores of the Moira River. We were especially interested in the great pipes out of which Belleville's sewage poured in robust and unfettered spate. Hundreds and hundreds of suckers used to gather below each outlet to gorge on this bounty and we would try to catch them — not, I hasten to add, for the pot but just for the fun of it. The suckers were picky eaters and hard to hook but we could at least always count on fishing out a few condoms.

We knew these were not balloons but were used in some obscure, not-to-be-talked-about way by our elders. I don't remember which of us had the inspiration, but one day we took some back to the Mowat apartment. We filled a couple at the kitchen tap with as much water as they would hold — perhaps a gallon — then carried them into my parents' bedroom at the front of the library. A bay window opened directly over the front steps and from this vantage point the captain and crew of the airship R-100 dropped their bombs.

The target turned out to be one of my father's friends, Eardlie Wilmott, a dashing young fellow who owned the local Ford dealership. Minutes later Geordie and I were trying to explain ourselves to Angus and Eardlie, who were sufficiently amused not to punish us, except verbally. However, we were sternly warned never to breathe a word about the incident to anyone, and especially not to our mothers.

Geordie's mother was renowned for her sensibilities. She once wrote a note to my father asking him to restrain his son from the use of indecent language. The phrase of mine to which she had taken exception was "I'll bet my bottom buttons." Had she heard about the condom bombing, she would have instantly severed all contact between me and her children.

My mother was, thank God, much more open-minded. The first time she heard me say shit (shortly after I had been enrolled in my new school in Belleville), she did not scold me. Instead, she explained that this was "not a *nice* word used by *nice* people. So try not to use it, dear. But if you ever feel you simply *must*, you *could* say shite."

Not long afterwards I tried her version on an older boy at school. "Don't you give me none of your la-di-da stuff!" he snorted, and bopped me on the nose.

Helen was pregnant when we moved to Belleville and early in 1929 was whisked off to hospital for the delivery. This proved difficult and protracted and, in the end, she lost the baby who would have been my sister. One day during my mother's convalescence I visited her, bringing a bouquet of lilacs I had picked myself. To my perplexity and distress, this caused her to burst into tears. Only later did I learn that the lilacs had been in full and early bloom the week that I was born, and she had believed them to be a token of good luck. Helen continued trying to bear but, after two miscarriages, gave up the attempt to save me from being an only child.

While she was in hospital, I stayed with my maternal grandparents and their youngest son, Arthur, who had moved to Belleville after Hal Thomson took "early retirement" from The Bank in Port Arthur.

Hal was an ideal grandad. Humiliated by his fall from grace, he avoided adults but felt at ease with me. A rotund and rubicund little man, he possessed a sweet voice and loved to sing me old songs from his youth. They were terrific songs. One was about a brash young whale: "In the North Sea lived a whale...big of bone and large of tail" who encountered a large silver fish and tried to bully it out of the way. "'Just you make tracks,' cried the whale...then he lashed out with his tail." But "That fish was, indeedo... a naval torpedo...and Oh, and Oh, how that poor whale did blow!" Another had to do with the adventures of an Irish rover called O'Shea who was cast away in India and became advisor to a maharaja.

> *He wrote to his sweetheart*
> *In far off Dublin Bay,*
> *He wrote to his sweetheart,*
> *He wrote her just to say...*
> *Sure I've got rings on my fingers*
> *And bells upon my toes,*

Elephants to ride upon,
My little Irish Rose.
So come to your Nabob
On next St. Patrick's Day,
Be-e-e-e Mistress Mumbo Jumbo Jijaboo Jay O'Shea.

Grandmother Georgina Thomson was a horse of a different colour. She was rangy, long-nosed, and an energetic, skittish, and unpredictable maverick. When Hal's career ended, George (as she preferred to be called) took control of the failing family fortunes and thereafter ran the show. Nothing daunted her. When she was in her late seventies she learned to drive, bought a red roadster, and drove it to Florida. But back in 1929 she had little time to spare for children or grandchildren. Arthur (of whom she once said, with a disdain for dictionary meanings that was legendary, "Well, my dear, he was an afterbirth, you know")* largely fended for himself. In consequence he spent much time at our apartment or aboard *Stout Fella*.

Arthur had a diabolical streak, as in truth did all the Thomson boys.** Left to keep an eye on me while my parents

---

When George was puzzled as to what course to take, she would complain that she was in a terrible quarry, and when confused she would find herself all of a zither.
**Geddes, the eldest, once stripped brother Jack to the buff, painted him all over with blue house paint, then shut him up in the back of a horse-drawn hearse which was parked in the street waiting to receive a neighbour's corpse. Jack lost some skin and hair when the paint was removed with turpentine. He was probably lucky not to have died from lead poisoning.*

were off at a party, he would devise ways to scare me half to death. Once he told me a really gruesome story about rotting corpses in a cave. He then hustled me into bed, turned out the lights, and let me stew for ten minutes or so before thrusting the great woolly head of a mop through the open transom above my bedroom door while giving voice to hideous howls, snarls, and screams.

Next door to the library on the south was a somewhat ramshackle house sheltering a family of five daughters and their mother. The father, a missionary-surgeon, had died in Africa some years earlier. The daughters took an interest in me and I was flattered when they included me in their pastimes. Mostly these seemed to revolve around medical emergencies — with me as the patient. All the girls planned either to become nurses or doctors and I believe most of them did so. I met one in 1944 in Italy where she was serving as a nursing sister in a Canadian military hospital.

"You were a big help to us in Belleville, Farley," she told me. "We had most of Dad's old medical books to study but pictures aren't enough. We learned a lot from checking you over. You were sort of our own little male cadaver, if you know what I mean."

I didn't learn much from them in return but I did have one unforgettable experience in their house. I was daydreaming in a swing sofa on their porch when a sudden movement caught my eye. I looked up to see a huge spider in the centre of its web battling an equally enormous hornet. The duel was taking place only a hand's breadth away, and I felt myself being drawn directly into the world of the combatants and, in some inexplicable way, associated with them. I watched in wonderment as the velvet-clad black spider feinted warily, avoiding the dagger thrusts of the golden hornet's sting. Suddenly the spider drove its curved jaws — in the tips of which tiny jewels of liquid poison gleamed — into the back of the hornet's neck. At the same instant, the hornet curved its abdomen and buried its dagger in the spider's belly. The web, which had been shaking wildly, grew still as death overwhelmed the duellists. And I slowly emerged from something akin to a trance having, for the first time in my life, consciously entered into the world of the Others — that world which is so infinitely greater than the circumscribed world of Man.

From this time forward, non-human animals increasingly engrossed my interest, both in daily life and in my reading. I had already read my way through the anthropomorphic confections represented by the *Uncle Wiggly* series, Beatrix Potter, Mother Goose, Aesop's Fables, the Christopher Robin books, and *The Wind in the Willows*. Now I began seeking stronger meat. The Dr. Doolittle books held me for a time but I soon moved on to Kipling's *Just So Stories* and, in the realms of my imagination, became another Mowgli. I also read everything about the Others written by Ernest Thompson Seton and Charles G. D. Roberts, whose animal characters were not always shaped in the human form but were sometimes allowed to be themselves.

As an almost inevitable result of having parents who were book worms, and of being immersed in the library atmosphere, I was far in advance of my age as a reader.

Nothing in print was forbidden to me. When I was seven, I worked my way through a big, lavishly illustrated volume of *Gargantua and Pantagruel* and, though I must admit I did not understand much of the text, I certainly appreciated the marginal drawings of grotesque human beings engaging in things dear to the imagery of small boys, especially the representations of Rabelaisian farting and pissing.

My tastes ran the gamut from the *Boys' Own Annual* to privately printed erotic novels given to my father by book salesmen. I was especially taken by one called *Aphrodite* whose limpid and stylish prose, and frank but subtly executed illustrations, made most modern pornography, even that which is classed as "art," seem as crude and rude as graffiti in a public toilet.

When it came to books and reading, Angus gave me as much latitude as I could wish for, but when my behaviour trespassed against the honour code he could become an Old Testament kind of parent. Left alone in the apartment one rainy Saturday afternoon, I entertained myself by making pellets out of tightly rolled wads of cotton wool which I shot at flies through my tin pea-shooter. Being nearly weightless, the pellets proved ineffective until I hit on the idea of soaking them with my mother's perfume.

Why perfume? I cannot say. Water would have done as well but perhaps I thought the stricken flies would die happier this way. In any event, when my parents returned home I was accused of fooling about with Helen's toiletry. I denied the charge, which was idiotic because the scent of Attar of Roses hung around me like a cloud.

Helen cried a little at the wastage of her treasured perfume and I had a little cry myself after getting seven whacks with my father's razor strap on the palm of each hand. It didn't really hurt that much, but since this was the first time I had been strapped, the shock was considerable.

After the ordeal was over, Angus and I solemnly (if somewhat gingerly on my part) shook hands while he assured me the incident would now be forgotten. I understood that I had not been punished for taking the perfume so much as for having lied about taking it. Lying by others was a cardinal sin as far as Angus was concerned, although he was a skilled practitioner at the art himself, especially when he was engaged in concealing his affairs with other women from my mother.

During the Belleville interlude I also dabbled in theft. Angus had a habit of leaving small change on top of his bureau. Since I had no fixed allowance I took to pocketing the odd penny. When I had amassed five cents, I would buy a special kind of candy bar sold at a little shop on the route to school. It wasn't the candy I was after but the wrapping. This consisted of stiff, perforated cardboard designed to be converted into a little glider that would actually fly quite well. I acquired several of these bars but guilt tormented me and I was afraid either to eat the candy or to assemble the planes. I kept my hoard in a hole in the limestone retaining wall behind the library. Eventually the squirrels found them and that was that.

My parents' bedroom and possessions exercised an irresistible fascination for me. I pawed through all their belongings at one time or another and can vividly recall the cold chill when I found Angus's .45 Colt service revolver under a pile of his underwear. I showed it to Geordie Sobie. Neither of us had the guts to touch this sinister-looking object but I nerved myself to steal one of the bullets. Possession of this shiny, brass cylinder with its ugly, snub nose of grey lead gave me considerable prestige at school for a few days — until an older boy took it from me.

Although it may seem odd, that incident is one of the few distinct memories I retain of my years in public school. Could it have been that my teachers were such singularly colourless souls as to fail to make a mark? Or did I simply dislike the servitude of school so fervently that I was able to eliminate its echoes from my mind? Whatever, my early school days remain as insubstantial in memory as a miasma.

Stout Fella continued to play a prominent part in my life. Because Belleville harbour was a filthy place, full of sewage and industrial refuse, Angus found a summer mooring for the little vessel at the mouth of Jones Creek, a few miles west of town. This became our summer home. When we were not off cruising, Angus commuted to work each day, leaving Helen to lie reading on deck in the sun, or belowdecks if it came on to rain, and me to wander about with the sons of a farm family who lived nearby.

These people were not really farmers. They had a cow and some chickens and lived in an old farmhouse, which was so dilapidated they dared not make use of its upstairs rooms. The several sons, all under fourteen, never seemed to wash, dressed like characters from *Huckleberry Finn*, and were free souls. I found them irresistible — but then all my life I have had a special affinity for social outcasts. They swore with casual abandon; drank homebrew bootlegged by their parents; trapped, shot, and fished with complete disregard for the law, and were, in fact, a law unto themselves. I suspect that my enduring affection for anarchy owes much to the time I spent in their company.

My friends did not all come from beyond the pale. After Geordie Sobie was finally forbidden to play with me because of the perceived threat to his morals, Alan Evans became my best friend. Alan and his widowed mother, Alice, a slim and faery lady much admired by Angus, lived with her father on a decrepit estate on the edge of town. It included a huge and rambling old house fast going to wrack and ruin, acres of overgrown lawns and gardens, and, best of all, an uninhabited gate-house.

Alan's grandfather was a mining engineer who had a dark and gloomy laboratory filled with equipment for metal assaying. This was a place apart — full of inscrutable machines, retorts, furnaces, piles of ore samples and drill cores, all in a state of mysterious confusion. To Alan and me it was the sorcerer's workshop. Although it was forbidden territory, we used to sneak in and root around. Once we came upon a small slug of gold, gleaming dully in the bottom of a retort. Captain Kidd's treasure could hardly have thrilled us more.

In the early summer of 1929 the adult world was awash in hectic talk of fortunes being made on the stock markets.

The get-rich-easy mood must have infected Alan and me for we too decided to make a lot of money, with which to finance exploring expeditions to the far corners of the earth. Our first enterprise was manufacturing raspberry cordial from a gone-to-wild raspberry plantation on the grounds. This was not a commercial success. We made gallons of the stuff but drank more of it than we sold.

Our headquarters was the abandoned gate-house, full of wrecked furniture, dust, bats, spiders, and mice. One day inspiration came to me. We would trap the mice, skin them, and make a fortune selling mouse fur. The idea originated from my having read about moleskin smoking jackets. If moles — why not mice? We borrowed several mousetraps from the big house and eventually caught a mouse. But our attempts to skin it, using bits of broken glass for knives, resulted only in a small, bloody mess of fur and flesh, and two cut fingers.

Next we decided to start a rabbit ranch. Pooling our scanty funds we managed to buy a pregnant doe, whom we named Buffy. This project involved Alan's mother since we planned to keep our stock in a more-or-less abandoned greenhouse on the Evanses' estate. Alice Evans appealed to Angus to head us off. He was then attending a library course in Toronto, from whence he sent me the following:

> *Since you've formed the habit*
> *Of keeping a rabbit,*
> *Please follow these orders,*
> *For rabbits as boarders*
> *Are no joke in the least for Sweet Alice.*

See Arthur and worry
Him til he does hurry
To build up a hutch
For fair Buffy, as such
Will be better by far for Sweet Alice.

Take it out by the barn
On the Ketcheson farm,
Where you'll have to buy it
Some food for its diet,
And relieve of her worry Sweet Alice.

Now do not forget it,
Or else you'll regret it,
Because it will die,
And your mother and I
Will be weeping, and so will Sweet Alice.

In the end Buffy was eaten by a cat, so this scheme too came to a dead end.

Alan then suggested we turn the gate-house into a hotel. We passed the word around that we would be happy to receive guests at two cents a head, with raspberry cordial thrown in. For a while there were no takers, but the word spread and one afternoon guests arrived. They were two teenage couples from the wrong side of town and they were not interested in our raspberry cordial. They *were* interested in the beds in the upstairs rooms, raddled with mildew and mouse nests as these were. Alan and I were turfed out of

our own hostelry with the emphatic adjuration not to
return or to tell anyone what was going on or "you'll get
your little asses kicked up to your ears!"

In the financially euphoric spring of 1929, Angus had felt
flush enough to do away with old, black Henry and buy a
car more nearly befitting his rising status. He chose a pea-
green Model A Ford roadster equipped with a folding canvas top and rumble seat. He named this jaunty vehicle
Eardlie, in gratitude to the dealer, Eardlie Wilmott, who
had given Angus a bargain on it out of admiration for my
mother.

Helen had always had many beaux and continued to
attract men long after her marriage. Although I am certain
she never succumbed to temptation, she enjoyed male
admiration and was not above a little flirtation. This did not
bother my father; in fact, he may have felt it gave him
licence. With sporty little Eardlie to rip around in, he now
became even more the dashing caballero and began exploit-
ing his attractiveness to women to the full.

Either his successes made him careless, or perhaps the
stories he told my mother to cover his tracks simply became
too outrageous for her to accept. Whatever. The day came
when she charged him with being unfaithful to her.

Both she and Angus thought I was out of the apartment, whereas I was reading in a corner of my room when
the ruckus began. Since I had never known my parents to
quarrel openly, I did not know what to make of the raised

voices coming from the kitchen. Curious rather than alarmed, I padded into the hall, from whence I could see them.

Angus was squatting in front of the open door of our ice-box, balancing on his toes as he reached into it for a bottle of beer. Helen was standing right behind him brandishing both fists in the air. As I watched, fascinated, she drew back her right foot and, for the first and last time in their life together, struck my father. She kicked him so hard in the rump that he tumbled into the ice-box, hitting his forehead with sufficient force to make it bleed — though only a little.

Helen was appalled by what she had done. Her hands flew to her face, then she said, almost dreamily, "Oh, Angus. Oh, Angus," and crumpled to the floor in a classic faint.

According to everything I have since read about the effects of such incidents on children, I should have been horrified, terrified, or traumatized. The truth is that I was excited by this little drama which sent a delightful shiver down my spine. Had I been older I might have been tempted to applaud. But I knew it would be a mistake to reveal my presence, so I padded silently back to my room where I remained invisible while my mother and father patched up their quarrel.

For weeks afterwards they were extremely kind to one another and to me. And Angus was much more careful in how he conducted his affairs. Thirty years were to elapse before Helen would find him out again.

My parents became very active in Belleville's "younger set" this year. Membership in the Bay of Quinte Yacht and Country Club was de rigueur for "gay young things" (homosexuals had not yet hijacked the word "gay") and there they sailed and danced the summer days and nights away.

Stout Fella was often moored to the club wharf on weekends so I saw and heard a good deal of what went on. I delighted in the music of the dance bands, the frivolity of young men in white ducks and young women in flapper dresses, and the general air of carnival. I even liked the smell of Martinis that used to hang like a mist in *Stout Fella*'s cramped little cabin. The adults around me were a merry lot, all unaware of the darkness that was about to overshadow the lives of many of them. They were happy and carefree, and so were we, their children. It was the last fling of the crazy twenties, and an idyllic time to be alive.

FIVE

In May of 1930, a few days after my ninth birthday, Angus travelled to Windsor, Ontario, to be interviewed for the job of chief librarian. He made the journey with no great expectations. Canada was then reeling from the shock of the worst financial collapse in recorded history, which had shaken the industrialized world in the autumn of 1929. Following the Crash, the mood had become one of uncertainty, apprehension, and retrenchment. Those who had jobs thought themselves lucky and neither expected nor sought promotions. Angus expected the trip to Windsor to be a pointless if pleasant little jaunt for he didn't believe he had a chance.

Neither did my mother, until he called long distance (something rarely done in those days) to tell her he had been offered the job, and had accepted.

Although Belleville styled itself a city, it was no more than a county town. Windsor was something else. When Eardlie drove into its outskirts late in August 1930, I found myself entering a different and intimidating world.

Windsor called itself the Border City. Together with several satellite communities, it encrusted the eastern shore of the St. Clair River which marks the boundary between Canada and the United States. Just across the river lay the vast, smoky sprawl of Detroit, Michigan, the Motor City, where nearly half the automobiles that were already dominating the lives of North Americans had been or were being built. A lot of Windsor men worked in the Detroit auto plants and chauvinists on the U. S. side of the river liked to refer to the two communities as the Twin Cities, implying (and even assuming) that Windsor was no more than a northern suburb of Detroit.

This is not the way we felt about it. Our industrialist overlords and traitorous political collaborators had not yet succeeded in subverting our conviction that Canada was a nation in its own right. As far as my family was concerned, Detroit and the whole of the United States of America were alien ground and, no matter what similarities its inhabitants might have to us, they were foreigners.

Even though I was only a child at the time, the experience of living on the borders of a foreign nation which so obviously believed in its Manifest Destiny as the eventual master of the continent helped instil in me the fervent nationalism I still proudly maintain.

Our new home was a ground-floor apartment in a brick four-plex on Victoria Avenue, only a fifteen-minute walk from downtown, and the public library where Angus now lorded it over a staff of six.

I was enrolled at Victoria School, which was three times the size of the one I had attended in Belleville. Because I was a new boy and "a little squirt," I became an instant target for a number of boys bigger and tougher if not meaner than myself. This was something I had not previously experienced but, since I was fleet of foot, I wasn't often caught. Nevertheless the humiliation of being frequently on the run did not endear Victoria to me.

Nor was this the only humiliation I endured. One of my few surviving memories of the place concerns a Home and School entertainment devised by a woman teacher who was insanely enamoured of the stage. Three other small boys and I were conscripted to perform an utterly idiotic, simpering version of a "folk dance" opposite four little girls. By the time the practice sessions ended, we boys on the one hand and the girls on the other had come to detest each other with a virulence extraordinary in ones so young. But this was nothing to the detestation I felt for our merciless taskmistress. By the time we actually performed our horrid little travesty, I hated that woman so much I believe I could have slipped a stiletto between her ribs, had I known what a stiletto was.

Boys facing girls, we pranced towards our partners, then backed away while shrilling a terminally stupid song, some verses of which are indelibly inscribed in memory: "Pray what is your intensir . . . intensir . . . intensir? Pray what is your intensir . . . with a ransom-tansom-tizza-matee?"

whined the girls. To which we boys replied with ill-concealed loathing: "Our tension is to marry . . . to marry . . . to marry. Our tension is to marry . . . with a ransom-tansom-tizza-matee." For weeks afterwards, this line was iterated and reiterated by older kids in the school yard whenever any of us chorus boys appeared, and queries about how we were going to celebrate our wedding night rang in our ears like hellish carillons.

The only good thing about that event was that it brought Hughie Cowan and me together. He was one of my fellow performers, and our shared suffering made us friends.

Hughie's parents were Scots immigrants. His father, a cabinet maker in the old country, had set himself up as a house builder in Windsor and had done fairly well during the post-war boom years. When the stock market collapse initiated the economic catastrophe which would eventually engulf the majority of working-class and even middle-class Canadians, Hughie's father fell early victim. His little business failed and thereafter he could not find sufficient work as a carpenter or as anything else to earn more than the most meagre living.

The Cowan family was large and so suffered severely from the Great Depression, yet its members remained as open-hearted and open-handed as only people of adversity can be. I was always welcome in their crowded little house where Mrs. Cowan would urge me to eat more than my share of whatever food they had. Nor would I be allowed to go home empty-handed. If there was nothing better I would at least be given an apple to tide me on my way.

The Cowans owned a battered 1924 Dodge touring car and, when they could afford a few gallons of gas, they would drive out into the bountiful agricultural lands of Essex County to barter with the farmers for garden truck and fruit. I was sometimes invited to go along and I remember those foraging excursions as times of shared gaiety which belied the grimness of their underlying purpose.

Hughie and I spent much of our spare time exploring the river bank, factory dumps, and big patches of "waste" land which had been cleared and surveyed by speculators ("developers," they now call themselves) during the boom years, but had since been abandoned and were now returning to wilderness. Rabbits, foxes, raccoons, and other wild creatures were recolonizing these regions although they had to share the space with unemployed and homeless men who had nowhere else to go. These destitute humans lived in what were beginning to be called hobo jungles, which Hughie and I often visited. We were always offered something to eat or drink, even if it was only a spoonful of beans or a mouthful of tea, and we were never offered any harm.

Once a jobless young man from Alberta entertained us by playing his accordion and singing songs about "life on the road," which meant life riding the rods on the railroad. The songs dealt with a side of life far outside my experience. The refrain of one of them ran something like this:

Oh, the rails they is made of tempered steel,
And so's the hearts of the Bosses,

But the rails ain't never half so hard
As their hearts when they cuts their losses.

I didn't understand this until Hughie's father explained it to me.

"It means, laddie, that when things get tough the owners cut off the working stiffs, ye ken?"

It was an early but salutary lesson for me in the way the capitalist economy works.

Considering the limitations imposed on ten- or twelve-year-olds today, our parents accorded us an enormous degree of freedom. Nobody seemed overly concerned about where we might go or what we might be doing. Years later I asked my father whether he and Helen had worried about our tendencies to wander far afield.

"Certainly we worried. The way a mother cat does about a kitten that wanders off after a chipmunk. But we felt that keeping you in a nice safe cage would leave you with only the vaguest and perhaps the wrongest ideas of what life was really about. Chances have to be took even by the young."

My reading had now taken me into James Fenimore Cooper's *The Last of the Mohicans*, and other tales of the Red Men. Indian lore fascinated me but it was not until I fell under the spell of Ernest Thompson Seton's *Two Little Savages* that I decided to become an Indian myself. I enlisted Hughie and we formed a tribe of two. We tried to emulate Jan, the hero of *Two Little Savages*, in learning and in practising Indian ways.

In the late spring of 1931 the Mowats had discovered Point Pelee, the most southerly point in Canada. Here, on the shore of Lake Erie, was a relatively untouched wilderness of forest, sandy beaches, dunes, and marshes. My family visited Pelee often thereafter, usually with Hughie in tow, and we two little savages were allowed to camp out in our own home-made wigwam to savour life *au naturel*. It is true that my parents camped within hailing distance but they made a point of keeping out of our sight and, as far as possible, out of our Indian lives. When we were in need of food (which was frequently) or of reassurance (as when a thunderstorm came roaring in off the lake), we went to *them*.

One July day the tribe went hunting. For an hour we slipped through the forest glades, soundless as shadows, in pursuit of the elusive moose. We found none because there were none on Point Pelee. Tired and hot, we eventually decided to have a swim but were outraged to find our favourite strip of sandy shore pre-empted by invading white men who had driven their covered wagon (a big Buick) along the hard-packed beach and were noisily setting up an encampment on *our tribal land*! To make matters worse, the licence plates on the covered wagon told us the invaders were Americans.

This was not to be tolerated. Wiggling through the dunes and taking cover behind tufts of tussock grass, we stalked the unsuspecting pale-faces. When we were within range, we let fly two blunt-headed arrows at their big tent.

This was not mere childish posturing. We were both practised bowmen and, moreover, our equipment had been designed and its construction overseen by Angus. An enraged howl from within the tent announced that we had made a point. We swiftly withdrew over the dunes into the woods and headed back to our lodge, well satisfied that we had struck a telling blow in defence of our native land.

That evening a pair of policemen appeared at my parents' tent, enquiring about the presence of archers in the area. It appeared that a stray arrow had penetrated the tent of some tourists and had struck one of them in the ribs. No real harm had been done but the Americans had departed breathing fire and brimstone. Which, the policemen solemnly noted, was bad for the tourist business.

My parents kept their peace, so the police departed none the wiser. Then it was Hughie's and my turn to be interrogated. I knew that a flat denial would only get me a licking for having told a falsehood, so I modified the lie and explained that we had shot at a squirrel in a tree top near the shore without knowing, until too late, that there were people on the beach. *American* people, I emphasized. It was probably this attention to detail which saved our hides. We escaped with a stern lecture but were forbidden the use of our weapons for the rest of that stay at Pelee.

———————————

Not all our family excursions were local. My mother's parents had a summer cottage near Danford Lake in the Gatineau Hills north-east of Ottawa, and Helen and I joined them there for a month during the summer of 1932.

I loved the place. The cottage was constructed of rough-cut white pine and was pungent with the scent of turpentine. It was not a city home transported into the country, as are so many "cottages" of today; it was a place not only in but *of* the wilderness.

Ceilings and roof were one, consisting of a layer of tarpaper over bare boards through which one could hear, unhampered, the drumming of the rain, the skirling of the wind during a storm, or the clatter of birds' feet on the ridge-pole in a calm summer dawn.

There was no running water (therefore no obligatory baths for the likes of me). We carried our water up from the lake in buckets and it was not only fit to drink but tasted deliciously of balsam. There was no electricity but there *was* an ice-box, into whose maw we daily dumped a fifty-pound chunk.

The kitchen contained a cast-iron range which burned billets of hardwood and, for cooking during the heat of summer, there was a two-burner kerosene (coal-oil) stove. On the slopes overlooking the lake stood a marvellous out-house through whose open door one could watch in comfort and seclusion the comings and goings of loons, ducks, and herons.

The mellow light of the well-named Aladdin lamps made the plank walls of the cottage glow golden at night, and enabled me to lie awake on my iron cot reading until all hours, engrossed in books about adventures in far places, Indian lore, or the lives of other animals.

On a sandy little beach below the cabin lay an old punt, square of bow and stern and heavy as lead, in which I could row wherever I chose amongst bays and coves still fringed by virgin forests. And there was not a single outboard engine or, indeed, any engine at all on the water with me.

There were no other children to play with but I did not miss their presence. I was by no means alone, or lonely, because of the plethora of other creatures including deer, beaver, squirrels, skunks, and a family of otters.

Once I spotted a young black bear snuffling along a beach and rowed quietly to within a few dozen yards of him before he saw me. We looked at each other for what seemed like ages. I was thinking about my dream bear at Bingen and I wondered how this one would look in a checked cap. The idea set me giggling. The bear cocked his head quizzically for a moment then lumbered off into the woods as if, perhaps, doubting my sanity.

There was more than enough to do. If I felt so inclined I could catch pike and bass and take them home as contributions to the dinner table. There were berries to be picked. There were little red-bellied snakes and wood frogs to be caught and secreted in tins and jars under my bed. At no great distance was a little backwoods farm whose house and barn were made of rough-hewn logs. The old couple who worked the place welcomed all comers. The farmer taught me how to use a Swede saw and an axe (well, it was only a hatchet), and told me wonderful stories of his earlier days as a logger, driving big timber down the Ottawa River.

That month at the cottage when I was eleven was no mere interlude in my life — it was a revelation. Thereafter the desire to become one with wilderness and its native inhabitants would grow ever stronger within me.

The Depression had little adverse effect upon the monied high rollers of Windsor (or anywhere else, for that matter). As chief librarian and therefore something of a cultural icon, Angus was favoured by the patronage of some of these, one of whom was the president of the Hiram Walker Distillery in nearby Walkerville. This man sometimes took us on elaborate party cruises in his luxurious yacht, which had a crew of seven, including a black bartender with whom I used to fish off the vessel's gilded stern when we lay at anchor. The bartender also used to make brandy Alexanders for me — without the brandy.

Angus gloried in these cruises, although in his Bay of Quinte days he had despised power yachts, contemptuously calling them "stink pots." Now, clad in white flannels, a blue blazer with gold buttons, and a jaunty yachting cap, he could easily have been mistaken for the film star Douglas Fairbanks, Jr., whom he not only resembled physically but whose swashbuckling, devil-may-care, life-of-the-party style he adopted as if it were his own.

Helen did not entirely share his pleasure in consorting with the rich.

"He would buy the fanciest yachting togs for himself but there never seemed to be money for me to dress so I could feel at ease amongst all those swells. Quite often I stayed home with you which, I am sure, didn't cramp your father's style at all."

Clothes or no clothes, Helen did not choose to miss an extravaganza held at the ornate mansion of the distiller to celebrate Christmas, 1931. I was taken along although it was definitely *not* a children's party. While the adults disported themselves downstairs, I was left alone in an enormous, chintz-filled sitting room on the second floor to entertain myself as best I could with a pile of silly games and a tray of cake and cookies.

The games soon palled so I explored the room in search of things to do. The house had been built at a time when electricity was available only to the rich and, instead of the more-or-less foolproof outlets in later-day use, had been fitted with brass wall receptacles into which one inserted a solid brass plug through a little trapdoor. Never having seen anything like this before, I tried the effect of pushing my finger through one of the trapdoors. The shock knocked me unconscious and half-way across the room, but did me no permanent damage except, perhaps, to introduce an element of caution into my future dealings with the leisured class.

My allowance during the Windsor years was five cents a week, and Hughie got none at all. Having seen how the rich lived, I concluded this situation wasn't good enough and looked around for ways to supplement our incomes.

The public library was shaded by great sycamore trees, and when these fecund giants dropped their seeds the surrounding lawns and sidewalks were deeply littered with them. Hughie and I gathered a bushel of the seeds which we packaged in Windsor Public Library envelopes and hawked from house to house. Our sales pitch was that the seeds would produce statuesque sycamores a hundred feet in height...which they might well have done, given an equal number of years. Unfortunately Windsorites were not interested in ensuring a forested future, so our sales were minuscule. Even a letter of mine published in the Windsor *Border City Star* and containing this trenchant declaration: "Sicamores [sic] are the rarest and beautifulest trees I have ever seen in my life and should be spread," failed to initiate a sycamore-seed-buying spree. We gave up on sycamores in favour of another venture.

People with cupboards full of unwanted magazines sometimes could not bring themselves to consign these to the garbage. Instead they donated them to the library with the result that the cellars were full of copies of the *National Geographic* magazine. Nobody seemed to want these so I asked Angus if I could have them. He was delighted to get rid of them, and Hughie and I spent the next two Saturdays hauling hundreds of *Geographics* in our Express wagons to the basement of my home. My plan was to set up a business supplying missing issues at five cents apiece to people who wanted to own a complete set.

It turned out there were all too few such; but Hughie and I at least increased our knowledge of the world from thumbing through our stock. We even learned a little about sex from avidly scanning pictures of bare-bosomed African and South Seas women.

The failure of these enterprises tended to sour me on a genre of books which was then considered essential to the proper rearing of young North American males — books like the Hardy Boys, the Tom Swift series, and especially the Horatio Alger books, whose dirt-poor but dogged young heroes made their ways to fame and fortune by devoutly pursuing their own selfish interests. Hereafter I devoted *my* attention to such nonconformist tales as *Peck's Bad Boy, Huckleberry Finn*, and *Penrod and Sam*.

It was in the spirit of the latter that one summer Hughie and I organized a circus in my back yard. Hughie was the acrobat, and a good one too, although one afternoon he gave the audience more than their money's worth by trying to do a handstand on top of our back fence. He crashed to the ground with such violence that the several little girls who made up our audience screamed their heads off.

I was the Master of Ceremonies, but also did my bit as an entertainer. Shrouding myself in a sheet upon which we had painted the outline of a skeleton, with a pillowcase painted to look like a skull pulled over my head, I played a ghost. I nearly became a real one when, blinded by my disguise, I ran full tilt into a concrete pillar, knocking myself out and leaving a permanent depression in my left temple.

We charged a one-cent admission fee and made a profit of fourteen cents, which we spent on two pomegranates (five cents each) and four one-cent, round, black candies called nigger balls.

That was the last spring Hughie and I shared together. Faced with the humiliating prospect of having to "go on relief" in order to feed his family, Hughie's father chose the desperate alternative of becoming a homesteader in northern British Columbia. I remember how Hughie put it, echoing his father: "We'll not become beggars. We'll starve first."

It was late June when they departed. Four children and three adults (one of Hughie's uncles accompanied them) crowded with all their baggage into their ramshackle old touring car. Camping out along the way, it took them six weeks to reach their destination — a one-hundred-and-sixty-acre plot of primaeval forest. There, with axe and shovel, they began making a new life.

Hughie and I wrote to each other off and on until the 1950s when we finally lost touch. In the autumn of 1992, his sister Margaret appeared at a store in New Westminster, British Columbia, where I was autographing books, to tell me Hughie was alive and well and still living on part of the original family homestead. There I will visit him one day, God willing, and tell him how much I missed him when he left me behind in Windsor in the time of our childhood.

SIX

My parents had given away our two cats before we moved to Windsor, but soon after we settled into our new home I found a scrawny stray kitten and brought it home. Angus named her Miss Carter after our landlady, whom he did not like. He used to stand on our front steps at night loudly calling our Miss Carter by name, adding: "Come home, you little tramp!", while the other one, who lived only a few doors down the street, presumably endured acute embarrassment.

Other animals joined us. One was Limpopo, a Florida alligator brought to me by my Uncle Geddes, whose fiendish sense of humour as a youth had suffered no diminution with the years. When Limpopo arrived, he was a starving, six-inch weakling, but he grew apace on a diet of hard-boiled eggs, minnows, and tadpoles until, before we left Windsor, he was two feet long and could glutch down a six-inch sunfish in a couple of gulps.

Although not an affectionate pet, he never tried to eat my fingers, as he might well have done. When we eventually left Windsor, I was told he could not accompany us. I tried to give him away but found no takers. In the end, I took Limpopo down to the banks of the St. Clair River and fed him half a dozen wieners as a parting present before sadly releasing him into the murky waters. Years later I read a news account of a six-foot alligator having been found in the Detroit sewer system. I like to think it was Limpopo.

Jitters was the most endearing of my non-human companions. During one of our visits to Point Pelee I had climbed a pine tree containing what I took to be a crow's nest. It turned out to be the home of a family of black squirrels. I hauled a half-grown young one out of the ball of twigs, stuffed it inside my shirt, and climbed down, much pleased with my acquisition. I was less pleased when I found myself crawling with squirrel lice.

Nevertheless, I was allowed to keep my find and bring it home, where both of us were deloused. Angus then built a clever cage with two compartments and sliding doors, by means of which the squirrel could be shut up in one half of the cage while the other half was being cleaned. This arrangement soon became unnecessary. After a few days Helen began to feel so sorry for the little creature, who spent most of his time clinging to the wire screening and "jittering" piteously, that she suggested we might try letting him out for a little while. This I hastened to do, with the consequence that Jitters soon became a full member of the family, enjoying the freedom of the household.

It was truly remarkable how he fitted himself into our lives — "imprinting," as Konrad Lorenz might have described it. He house-trained himself within a week, returning to his cage to defecate or urinate. He quickly learned not to steal food but would sit up on his hind legs and beg so prettily for what he wanted that he was seldom refused. Amazingly, he did not chew our belongings or burrow into the upholstery.

His one departure from otherwise almost impeccable behaviour was the merciless way he tormented Miss Carter. Being skinny and undersized, Miss Carter had an inferiority complex to begin with. Jitters made it worse. She spent hours hiding in dark closets rather than endure his gibbering verbal assaults which were often accompanied by a rain of small objects ranging from cigarette butts to walnut shells hurled down upon her from some high point of vantage.

Jitters loved to entertain visitors by racing around the edges of the living-room floor until he had gained maximum velocity whereupon, like a circus motorcyclist, he would begin spinning around the walls, spiralling higher and higher until he reached the picture rail up under the ceiling.

This stunt proved to be his nemesis. One spring day in 1932 he was racing around the *outside* walls of the apartment building (which were of rough-textured brick) for the entertainment of a group of my friends, when he lost his grip. I saw him shoot off into space from a corner up under the eaves. His little body described a swift parabola over the street then thudded to the pavement. He died in my hands a few minutes later. All of us, except Miss Carter, wept for him. If he had lived, he would have been the first black squirrel ever to trek as far west as Saskatoon, and heaven only knows what effect that might have had upon the world's equilibrium.

Pets were by no means my only connection with the Others during my years in Windsor. Probably as a consequence of my earlier adventures with bees, wasps, and spiders, I developed a curiosity about insects in general, and moths and butterflies in particular. Collecting moths and butterflies became a major preoccupation. During the warm seasons I seldom went anywhere without my butterfly net. So equipped, I became a figure of ridicule to some of my contemporaries and I think even Hughie found it somewhat embarrassing to be seen with me.

I learned how to "set" the insects on wooden forms so they would dry in the right position to be mounted on long pins in the bottoms of old cigar boxes. Angus made the frame for my net and Helen covered it with fine muslin. Angus also got me a killing bottle, which consisted of a glass pint sealer into the bottom of which had been poured a mixture of plaster of Paris and cyanide of potassium covered with a layer of cotton wool. The cyanide fumes would almost instantly kill any insect placed in the bottle and, I now realize, could have killed me had I taken a few deep breaths of the fumes myself. What I had was, in effect, a small-scale model of Hitler's final solution to the problem of non-Aryans. It amazes me to think how readily children of my day could come by such lethal devices. At the age of thirteen I went all on my own to a drugstore where I asked for and was sold a cyanide killing bottle made up to order, no questions asked.

Although I collected butterflies with enthusiasm, it was moths that truly fascinated me. I spent countless summer evenings patrolling beneath the street lights on Victoria Avenue to which the mysterious luna moths and the great cecropia and polyphemus moths with their five-inch wing spans were sometimes attracted. I can still smell the fragrance of those summer nights and feel the wild exhilaration of capturing a rare specimen. Although I would not now commit such atrocities against some of the most beautiful creatures extant, I cannot honestly censor the boy-who-was for what he did then.

In December of 1931 I acquired my first dog. My parents and I had paid a weekend visit to one of my mother's former Trenton friends who was living in Cleveland, Ohio. This lady owned a pure-bred Boston Bull which had given birth to a litter some six weeks before our visit. The father of the pups had been an unidentified travelling dog (possibly a dachshund) and the pups were an embarrassment to their owner. I immediately fell in love with them and pleaded to be allowed to take one home. Angus said he thought there were restrictions about importing dogs into Canada, whereupon Helen volunteered to smuggle a pup across the border, concealed inside her coat.

All went well on our return journey until the customs inspector came up to our car at the Canadian end of the Ambassador Bridge, leaned down, and asked through the open window if we had anything to declare. At this my mother drew her coat so tightly around herself that the pup squealed shrilly in protest.

Wedged between my parents on Eardlie's narrow seat, I was sure that all was lost, but I had not reckoned with Helen's ability to deal with the unforeseen. Without a second's hesitation, she threw back her head and began to squeal with high-pitched abandon. Angus, quick on the uptake as usual, cried out, "My God! She's getting hysterical. The doctor said it might happen. I've got to get her home!"

The customs officer hurriedly waved us on for no man likes to be involved with an hysterical woman. We were all three in near hysterics as we turned down Victoria Avenue, home again.

Billy, as I named the puppy, was ill-fated. In the spring of 1932, he was so badly mauled by a neighbour's Alsatian that he had to be destroyed. Jitters did not long survive him. Then, in June, Miss Carter disappeared. I suspected the Alsatian, who was also a notorious cat killer, and, in my desolation at having lost so many of my friends in so short a space of time, I contemplated borrowing my father's revolver and taking vengeance upon the dog. I lacked the courage. I tried to vent my feelings by writing an epic poem celebrating my departed companions and excoriating the villain. It has not survived but I can remember a couplet referring to the Alsatian:

I'd like to choke him full of mud,
And drown him in his own foul blood.

I am not sure when I started writing verse but the dreadful seed took root after the Christmas season of 1930. For reasons which were never clear to me, I was bundled off to spend the holidays with my maternal grandparents in Belleville. I was *told* this was because my parents both had influenza, but I now suspect it was because there had been a major rupture in their relationship, perhaps due to another episode in my father's long string of infidelities. If this were indeed the case, then the poem, "Daddy's Dilemma," which he sent to me in Belleville is demonstrative of his ability to set out false scents. Here is a truncated version of it.

Preserved by a Princess
or
Daddy's Dilemma

Your Mother still is stretched upon her bed,
*   With paunchy pillows pushed beneath her head,*
While our Miss Carter doth disport herself
*   On counterpane, like dizzy, dev'lish elf.*

Her name to Persian Princess has been changed,
*   (P.P. for short) and she has caught the mange,*
From sleeping, while in Oakville, with big Joe,
*   (They all have mange in Oakville, as you know.)*

But, as I said, your Mother had a break,
*   Short-circuited her nerves, and like a snake*
That winds and writhes and ties itself in knots,

She her sweet self all twiny twisted got.
(Don't try to say this fast — it's better not.)

The doctor came. His face was ashen grey;
  "In bed," shouts he, "a fortnight you will stay!
Some little medicine I'll give to you;
  For food, I'll let you gnaw a giblet stew."

Then came a nurse, in cap and apron white,
  And gave P.P. and me an awful fright.
She chased us to the basement dank and dark,
  Where P.P. wailed and mewed, whilst I did bark.

There P.P. lost her voice and 'gan to wheeze,
  And Daddy doubled up with cough and sneeze.
We both caught cold down in the dismal cellars,
  And sniffed and snuffed and snotted in our smellers.

But P.P. hit upon a clever plan,
  To fool the nurse and save her loving man.
She dashed upstairs, all streaked with dirt and dust,
  And told the nurse that she had seen a ghost.

The nurse came down and found me lying there,
  All stark and stiff upon the cellar bare.
"Good lack!" quoth she, "What makes you look so old?"
  "Fair dame," quoth I, "I fear I'm taking cold."

Her bleak eyes blazed with blue and bloody glee;
  She laughed, and sulphur cinders sprayed on me.

She screeched aloud, while from her raving mouth
  Red flames went twisting east, and north, and south.

"Gadzooks!" cried she, "I'll do whate'er I can,
  To heap more torments on you, dismal man.
I'll fill you full of gasoline, then I
  Will touch a match to you and watch you fry.

"I'll pluck your eyeballs out with furnace tongs,
  And tie your tongue in knots with red-hot thongs.
I'll hitch your entrails up to yonder beam,
  And throw live coals upon you till you scream!"

Wherewith this jolly lady seized an axe,
  Her aim, apparently, to smite some smacks
Upon my unprotected, aging head,
  And plash her plods in places where I bled.

But in that moment P.P. saved the breach,
  Rushed to my rescue and with fearsome screech
Leapt on that maiden's back, and strange
  As it may seem, implanted there the mange.

"Begone!" she cried. "Avaunt! Avast! Alack!
  Depart, thou mangy cat from off my back!"
Then down the street she pooched with pingy pace,
  And ne'er again we saw her flatsy face.

Your mother sends her love, to which, as well,
  She adds a bit for Arthur and for Hal.

I long to see you coming home again,
  So I can play with your electric train.

Be bad, my child, and let who will be silly,
  You will be both, I fear me, willy-nilly.
I now remain, your ever-loving poet,
  Angus McGill, begettor of Bunje Mowat.

This poem stimulated me to emulation and I became an inveterate scribbler of doggerel. It was an accomplishment which served me well during my remaining school years for not only did my rhymes win me kudos from my teachers but my ability to pillory opponents in verse gave me some protection against mine enemies.

One grey January afternoon in 1933 Angus brought home momentous news. He had been offered the job of chief librarian in Saskatoon, Saskatchewan, a place so distant from the Ontario experience that most people, if they had heard of it at all, thought it was some sort of geographic joke. Despite the fact that there would be no increase in salary, and the Depression was deepening daily, Angus was tremendously excited. Helen did not share his enthusiasm.

"Surely you wouldn't do such a thing," she cried. "It would be like taking us to Siberia or some dreadful place like that!"

In common with most easterners, she envisaged Saskatchewan as an alien and hostile world, a frigid, wind-swept wilderness in winter and a featureless and dreary expanse of dusty wheat fields in summer. She believed it was inhabited (barely) by peasants from central Europe who lived in sod huts, wore sheepskin clothes, ate black bread, and practised obscure religious rites.

She knew for a fact that the prairies were in the grip of a devastating drought, one of whose manifestations was cyclonic dust storms which whirled the topsoil off eroding farms and left their owners destitute. She had also gathered from the newspapers that the Depression was laying an even heavier hand on the prairies than on Ontario. All in all it was beyond her comprehension why anyone would want to go to Saskatchewan.

However, Angus was determined to make the move, and I backed him up. Our motives were essentially the same. We were men, and Adventure was calling.

My own image of Saskatchewan was of an enormous green plain rolling to an unimaginably distant horizon, inundated by black hordes of buffalo and inhabited by Indian tribes who rode their horses as if they were one with their steeds. My twelve-year-old's imagination assured me that the world of the Wild West was still alive.

Although I did not take part in the discussions which ensued between my parents, I was well aware of them. I knew by my mother's air of gloom and bouts of tears that she was fighting a losing battle. I was glad of it because, as each day passed, I wanted ever more desperately to go west. I closed my mind to her travail. It was not until many years later that she told me how she had felt at the time.

"You see, Farley, what Angus was asking me to do was give up not just the world I knew but most of the people who were part of my life. I would be cut off from all those I treasured most, after you and Angus. Perhaps it was very selfish of me but I wanted to stay where I felt I belonged. Going adventuring into the unknown does not attract most women, you know. It frightens most of us, or so it did me."

At the end of January, Angus tendered his resignation to the Windsor library board, to take effect on June 30. We were committed to one of the great adventures of my life.

SEVEN

One of Helen's most potent arguments in her struggle to dissuade Angus from going to Saskatoon was that he would be leaving behind the world of waters and boats which had always been so much a part of him. No more cruising on the bay. No more voyaging on Lake Ontario. Saskatchewan was a "dust bowl," a semi-desert, and Angus must surely despair of such a place.

The argument was sound but it did not take into account my father's capacity for self-delusion, with which he now set about manufacturing the illusion of a maritime world in the distant West.

To begin with, he determined that we would not make the journey by train, as sensible migrants of that era did. No, we would become as one with the early pioneers and head out in a covered wagon. Although (in order to enlist my mother's romantic instincts) he initially described the proposed vehicle as a "kind of gypsy caravan," what he actually had in mind was a *prairie schooner*, one which he would design, build, and pilot himself. As captain of his own ship, with Helen and me as crew, he proposed to make the passage to Saskatoon in nautical fashion.

Through the good graces of another of his high-flying friends, Angus got free use of a heated building at the Corby Distillery in Walkerville. Here, in early February, he began constructing his vessel.

There were those amongst my parents' acquaintances who thought and freely said that the choice of this locale for such an endeavour was inspired. "If Angus wasn't drunk when he got that crazy idea, he will be before he gets very far along with it," was the opinion of Alex Bradshaw, our neighbour in the next apartment. But Alex was wrong. Although my father drank whenever conditions were right he was no alcoholic. He was a man so dedicated to his dreams that not even the proximity of tens of thousands of gallons of whisky could seduce him from his purpose.

During the next six months he spent most of his weekends and holidays building a ship's cabin about eight feet wide and fifteen long, mounted on the four-wheeled frame of a Model T Ford truck. Uncompromisingly square both fore and aft, it had a cambered deck high enough to provide headroom for a tall man. Angus framed his vessel with steamed, white oak ribs and sheathed her with tongue-and-groove cedar planking, over which he stretched an outer skin of marine canvas. "Ought to be able to stand up to a

hurricane," opined one of those who came to see her growing. "Yep, but it'll take a locomotive to shift her," another
concluded. Angus kept his peace. He *knew*, with the assur-
ance of perfect faith, that Eardlie would be up to the task of
hauling our prairie schooner half-way across the continent.

I shared his faith and helped him at work as far as I
could — which wasn't very far. Although I loved fiddling
with tools and wood, I could not then and still can't measure things with anything like the accuracy required of a
craftsman. One day when I had made a cut half an inch
short in a piece of wood for the caravan's frame, he said to
me, quite unkindly, "Bunje, my lad, you are without doubt
the roughest carpenter one man ever told another about.
Why don't you take up knitting or finger-painting?"

Angus was an excellent carpenter, and he fitted out the
vessel's interior with skill and cunning. The trim little galley
boasted a small ice-box, a gasoline camp stove, and a tiny
sink. There were two main berths in the afterquarters. A
smaller, folding, pipe berth, slung across the stern, was my
rookery. Built-in bookshelves, lockers, a table, and a settee
completed the furnishings. The whole was made bright and
airy during daylight hours by six large ports, and at night by
two brass oil lamps set in gimbals. The windows — sorry,
ports — were fitted with red and white striped awnings
which could be demurely lowered when the vessel was at
anchor.

Angus painted his new vessel green and christened her
Rolling Home; but she was better known as Angus's Ark
which, being difficult to say, was shortened to the Ark.

On Saturday morning, August 5, 1933, the Ark set sail on her maiden voyage — a trial run, as it were — to Oakville, from whence we would take our eventual departure for Saskatchewan.

The omens were not propitious. When only a few miles on our way, the unwieldy vessel (which, because she had four wheels, tended to sheer wildly from side to side) escaped my father's control and ricocheted off a curb, knocking several of the wooden spokes out of a front wheel. Angus had to drive Eardlie back to Walkerville and search out a new wheel, leaving Helen and me to explain to a crowd of the curious what the Ark was and what she was supposed to be doing. The reaction was one of incredulity.

"She'll never make it!" said an onlooker. "Nope. You'd best haul her onto the nearest bit of ground, Missis, and plant some flowers out in front, and settle down right here."

Helen might have been content to do just that but Angus returned with a new wheel, repairs were effected, and we continued to our first destination, Port Stanley on Lake Erie. We moored for the night alongside a friend's cottage and I went happily off to swim in the lake, while Helen cautiously cooked our first meal *en passage*. This consisted of scrambled eggs on toast, coffee (milk for me), more toast, and honey. As time wore on, she became somewhat more adept at coping with the galley stove which, if not carefully watched, tended to flare up and incinerate the cook.

The captain wrote in his log next morning: "A pretty sleepless night. In the next cottage, a party from Detroit made merry until 5:00 a.m. and Farleigh was seasick during the night and vomited over the side of his bunk into mine."*

I was *not* seasick. It was simply that the excitement of our departure had got to my stomach which was notoriously "delicate" — a condition which my grandmother Thomson blamed on "all those soda biscuits and honey the poor lamb had to eat when he was small."

The following day we reached Oakville, where we remained for two weeks with Angus's parents. They were then in their seventies and gloomily viewed our departure for the Far West as a final separation. Grandfather Gill numbed his sorrow with the contents of a bottle of rye which was Angus's parting gift. Grandmother Mary withdrew to her bedroom after having frigidly stigmatized the removal as "more foolish nonsense of the sort that has distinguished the Mowat men for generations."

We "hauled anchor" on the morning of August 21. Angus had used some of the intervening time to adjust the tow bar so that the Ark no longer sheered about like an unbroken stallion on a slack tether. Nevertheless, she did not tamely follow Eardlie, and the Captain still had problems

---

During our final year in Windsor, I had become increasingly unhappy with a Christian name which the other kids inevitably altered to Fart-ley. When I complained about this to my father, he proposed to solve the problem by changing the spelling. This, he claimed, would take the curse off it. So I officially became Farleigh and, as far as he was concerned, the matter was settled. Of course this attempt to disguise the obvious had absolutely no effect.

with the helm. "Going through London we found the narrow streets and fool street-cars a distinct nuisance," he
noted angrily. I hesitate to think what the streetcar drivers
must have felt about us. Certainly we must have been a trial
to motorists on the open highways for they had to dawdle
along behind, sometimes for miles, before finding a stretch
where they might safely pass our lurching behemoth.

Because the rumble seat was packed full of luggage, I
began the voyage crowded between my parents on Eardlie's
narrow front seat. All of us soon grew dissatisfied with this
arrangement and, after a few days of querulous discomfort,
Angus asked me if I would like to ride inside the Ark itself.

What a question! Would I have liked to skipper the
Queen Mary? Would I have liked to pilot the *Graf Zeppelin*?

The upshot was that I travelled most of the way to
Saskatoon in command of a vehicle which adopted as many
guises as my imagination willed. One of these was a World
War I Vimy bomber. Crouched on my pipe berth in the
stern, I kept my Lewis gun (actually a Daisy air rifle) swinging
from side to side as I waited for pursuing Spads or Fokkers to
fly into my sights. I would insult pursuing enemy pilots with
such gestures of disdain as wagging my fingers in my ears,
sticking out my tongue and, yes, even thumbing my nose,
before pouring a burst of machine-gun fire into their vitals.

My parents were baffled by the hostility displayed by
some overtaking motorists who shook their fists at Eardlie,
yelled insults, and on one memorable occasion flung a hot-
dog with such accuracy that it splattered mustard all over
Rolling Home's bluff bows.

Angus would bare his teeth at such displays of incivility and fling pungent epithets back while Helen, who hated displays of raw emotion, cringed in the seat beside him.

In 1933 one could not drive east and west across Canada because no road yet spanned the great hump of granite and spruce forests north of Lake Superior. Consequently, we had to cross into Michigan in order to make our way westward. This we did by taking a ferry across the St. Clair River from Sarnia to Port Huron.

Since Eardlie's best speed never exceeded twenty-five miles an hour our progress was leisurely. On a good day we might run a hundred and twenty miles. We made fairly good time on pavement but gravel roads, which became the norm the farther west we went, were our bane. Poor Eardlie could not seem to get a good grip on gravel, and slithered and slid about with abandon. Nevertheless he was always game and the log is filled with entries attesting to his fortitude. "This day Eardlie hauled *Rolling Home* over a steady succession of fairly high hills on the way to Grand Rapids, and did it without even a wheeze or a cough, though he did drink an extra quart of oil." It is notable that Angus always referred to Eardlie as male, and in terms which more nearly applied to a horse than an automobile. But *Rolling Home* was *female*, as a ship must be. It seems not to have occurred to him that the idea of a horse hauling a ship across the continent was somewhat bizarre.

Many of the people we met along the way certainly thought we were a bit odd; yet, for the most part, they were kindly and well-disposed. On one occasion we parked the Ark in a municipal tourist camp but the day was too hot for Helen to do any cooking so we drove to a roadside café for dinner. At a cost of forty cents apiece, we had southern fried chicken with all the trimmings, and apple pie and ice cream. The proprietor was friendly, but too inquisitive for Angus's taste. He wanted to know where we had come from, where we were going and, in both instances, why.

"I told him," Angus noted, "that we were sailing a prairie schooner to the west for a cargo of buffalo robes. The fellow looked out the window at Eardlie sitting there with his top down and replied thoughtfully, 'Sun gets powerful strong in these parts. Lotsa folks been known to git sun stroke!'"

The route we were following required us to take a ferry across Lake Michigan, but when we arrived at the docks on the eastern side of the lake it was to find that *Rolling Home* was too high to clear the vessel's doors. We were told our only hope was to try loading her on a railroad ferry which sailed from Ludington, another port well to the north. It seemed a slim chance but the alternative — to drive all the way south around Lake Michigan through Chicago and its environs (inhabited mainly by Al Capone's ruffians, so we believed) was not attractive. We headed north.

The men servicing the huge railroad ferry, *Père Marquette*, were amused but helpful. "We *might* put that thing [the Ark] on a flat car and ship 'er over as cargo but then she'd be too high to go through *our* loading doors. Nope, that won't do. But we've got a train to load aboard and if there's room behind the caboose we might be able to roll that thing on too."

Which is what they did. Ten men manhandled the Ark
onto the rails and aboard the ferry where they lashed her tight
against the caboose with Eardlie nosing up to her stern. We
went on deck but Helen was concerned about our Ark.

"Whatever will the poor thing think? One minute she's
a caravan, then a prairie schooner, and now a freight car. I
do hope she doesn't get confused."

The crossing to the Wisconsin side of the lake took six
hours and was one long delight. I was especially thrilled
when the second engineer took Angus and me below and
showed us the engine room. I was barefoot, having lost one
shoe the day before, and Angus was in flannels.* We were
sights to behold when we emerged on deck again but what
a spectacle those huge steam engines were, all brass and
gleaming motion and spurts of vapour.

We then showed some of the crew through *Rolling
Home*, in return for which the first mate asked us to the pilot
house where we spent a fascinating hour amongst the radios,
compasses, and other instruments, and I was allowed to put
my hands on the great mahogany steering wheel. Later, at
dinner in the saloon, we met a couple from Regina, the capi-
tal of Saskatchewan, on their way home by train. They talked
"west" with my parents and told us a good deal about the
drought and the dust storms which awaited us.

Driving on from Manitowoc the next day, we reached
Lake Winnebago, near which we anchored for the night,

---

*Angus was always a natty dresser, even when embarked on a pioneering
voyage such as ours. I was then and have remained quite the opposite.*

and *Rolling Home* became the recipient of considerable attention and admiration from the inhabitants of the nearby town. I wonder now. Was our visit the seminal factor which would one day unleash thundering hordes of Winnebago motor-homes to prowl all over North America? I devoutly hope we were not responsible for that.

Saturday, August 27 was notable because we had a strong tail wind with whose help Eardlie occasionally got up to thirty miles per hour and ran off a record passage of one hundred and seventy miles, consuming fourteen gallons of gasoline and three quarts of oil in the doing. Reaching the town of Hudson on the St. Croix River just before dusk, we anchored in the local tourist park. This dispirited acre of burned-out grass offered a superb view of the river valley, and some of the dirtiest toilets we had yet encountered. I went looking for birds while Helen went shopping at a nearby general store and Angus picked up a twenty-pound block of ice.

In the Thirties every town and most villages had a public tourist park which provided, usually free-of-charge, outhouses, running water (cold), fireplaces, and sometimes firewood. These refuges were much used by migrants moving about the country in search of work. One such family was in the park when we arrived. It consisted of an aged, extraordinarily tall, thin, dirty, stockingless man; three boys from ten to fourteen in tattered overalls; two girls, quite comely in men's trousers; and a baby a few months old. What the relationship between them all was, heaven knows. They were travelling in a hopelessly dilapidated car,

towing a broken-down trailer in which they carried a tattered tent and their camping gear. The girls seemed to do
all the work while the old man slept. According to what
one of the boys told me, they originally hailed from Texas,
and had been on the road since March and were heading
north to hoe potatoes.

Although sympathetic with their plight, Angus thought
these people shiftless. Helen was a little frightened by them.
I found them interesting and became chummy with the
boys, who much admired my air rifle, but I thought them
greedy for the way they devoured a plateful of cookies
Helen offered them.

Much later, while reading Steinbeck's *Grapes of Wrath*, I
would remember the cookie incident with a pang of shame,
but at the time I had no comprehension of the miseries and
degradation to which that family and several millions like
them were being subjected. For them, the economic collapse of 1929 had not been a "depression" but a bottomless
pit into which they had been plunged with small hope of
escape. Most of the people with whom we shared the
municipal "tourist" parks were, in fact, Depression refugees,
desperately seeking work of whatever kind wherever they
could find it. Although most seemed a cut above the family
we met at Hudson, they were all enduring adversity of a
severity hardly credible to most of us today.

On September 1 we were approaching Fargo, North
Dakota, when, with astonishing abruptness, we found ourselves on the prairie. "Hell's bells!" cried Angus as we stared
across a world with no apparent horizon. "We're at sea!"

We headed due west, into the blue, and it was goodbye pavement, goodbye hills, goodbye trees and shade and sparkling brooks. We drew our first deep breath of prairie dust. Eardlie squared his shoulders and his engine took on a deeper hum. On every hand, threshing machines were at work. Straw stacks were burning, sending blue smoke plumes into a bluer sky. Horsemen trotted across vast reaches of virgin sod where cattle grazed. Gophers popped up and down on every hand and rattlesnakes slithered into the ditches. We passed lonely, treeless, unpainted houses from which ragged children poured out to gape and wave at our swaying green house on wheels. Late in the day we stopped on top of a little hill. As far as we could judge from the road map, we could see forty miles across the prairie in every direction. It was awe-inspiring, for it seemed to be a never-ending vista.

The fascination of it for me was intensified by the stupendous numbers and varieties of animals. There were no buffalo, but gophers (ground squirrels) of several species seemed to be everywhere. Ducks by the tens of thousands clustered noisily in the few ponds and lakes (sloughs,* we would learn to call them) that still held water. Huge hawks hung in the pale air or perched on telephone poles along the road, eyeing us balefully as we trundled by raising a cloud of dust behind us. Red-winged and yellow-headed blackbirds flowered like tropical exotics in roadside ditches; western meadowlarks sang loud and clear from the fence

*Slough is pronounced "slew."

posts, and coveys of partridges and prairie chickens shot out of the wheat stubble like miniature rockets.

We camped near Minot and, with my trusty Daisy in hand, I took my first walk on the prairie in palpitating fear/hope that I might meet a coyote. And I did. But he was dead — long dead and desiccated from sun and wind. His lips were drawn back over white teeth in the dry rictus of a snarl, and one hind leg was firmly clamped in the grip of a rusty steel trap.

The next day I saw a live coyote slip like an ochre shadow into the tumbleweed in a coulée, but the dead coyote looms larger in memory.

We were then driving north within a day's journey of Canada. The heat was fearful and the world burned brown. We passed a sign that read "swiming 21 mi," and our hopes rose. But when we got there it was to find a miserable little alkali lake and a ramshackle and seemingly abandoned dance hall set in the bleakest, most desolate situation that could be imagined. The lake was almost literally alive with thousands of mallard ducks which were not so much "swiming" as wading about in the muck. We did not disturb them.

On September 4 we crossed the border at Portal, North Dakota. Helen and even Angus were appalled by our first view of Saskatchewan. It looked like a desert in the making. Nothing green to be seen. Rough little valleys cut through low brown hills with not a drop of water in them. Here and there in the valley banks were the mole-like black holes where people had been digging for coal but we saw no people, no cattle, not even any gophers.

Seeking a tourist camp, we got lost and ended up in an abandoned village where gaunt, grey wooden buildings leaned against each other on an empty "main street." We headed for Estevan, the nearest town shown on our map but, before reaching it, encountered a few poplars still sporting some green leaves. There was a weathered farmhouse not far away so Angus went to ask permission to camp in the grove. Mr. and Mrs. Gent graciously gave it, but pressed us to park the Ark in their farmyard instead. Angus wrote of them:

"The Gents and their farm seemed to us rather a pathetic spectacle. They are English immigrants who have been 24 years here and raised a family, all of whom have had to leave the farm. The old couple — they were not really old although both looked it — were no farther ahead than when they landed in Canada. Yet, in spite of poor crops, drought and having to board coal-miners to stay alive, they remained cheerful and optimistic. Yes, there *had* been three years of arid drought, but next year it might rain. They *had* sowed 250 bushels of wheat this year, and harvested 500. But *next* year things might be better!

"They have two cows and a little milk route in Estevan, nine miles away. They sell a hog or two each year and some fowl, and so: 'We have managed to keep off Government relief, and that is something these days.'

"Because of the drought they have to cart their drinking water from Estevan. Washing water comes in a ditch from the nearest coal mine and is black as tar. They were most insistent that we stay with them next day, which was

Sunday, for a chicken dinner and seemed deeply disappointed when we had to refuse. Having heard me complain that
I did not like American tobacco, Mr. Gent pressed a pack-
age of Old Chum on me and would accept no payment. I
don't know why the Almighty couldn't let such folks have a
little rain occasionally."

Next morning Saskatchewan showed another face. We
woke in a chill grey dawn so cold we had to run the
Coleman stove to heat up the cabin. We washed in cold
coal water and then bade adieu to the kindly Gents and
rolled out across the prairies bound north and west for
Saskatoon.

But not before Mrs. Gent, cautioning me to secrecy,
had slipped a fifty-cent silver piece into my pocket. I did
not tell my parents about this until much later in the day.
They concluded that this may well have been the only
"cash money" the Gents had in hand but, after a great deal
of discussion, decided not to send it back for fear of mortal-
ly offending them. Angus kept the money "in trust" for me,
but not many weeks after taking over his new job, he began
shipping library books to the Gents, who had mentioned
that they could seldom find anything to read.

From this beginning, Angus eventually developed a
travelling library scheme by means of which the Saskatoon
Public Library circulated thousands of volumes to remote
parts of the province where people had no other access to
books. It was his contribution to easing the miseries of the
Depression and it was no mean one either. Before we left
Saskatoon, the library had accumulated a fat file of letters

from people who wrote that the books they had received had meant as much to them as food.

Angus and Helen remained in touch with the Gents for several years. I shall not forget them. They were of the enduring quality that distinguishes people of adversity. I was to find many more like them in Saskatchewan.

EIGHT

Saskatoon still lay some four hundred miles to the north-west. As we slowly made our way towards it over dusty clay and gravel roads under a brilliantly clear autumnal sky, the face of the land began to assume a friendlier guise. Although these prairies were also drought-stricken, they had not been so desperately ravaged as those in the south. The saffron-coloured wheat fields rolling away on every hand were at least yielding something in the way of crops to the men, horses, and machines that crawled across them. There were fewer abandoned farms. Although surrounded by glaring white pans of alkali, many of the larger sloughs still held central pools of water which were crowded with ducks. As we approached Regina, we began to pass small groves of aspens and poplars.*

These "bluffs," as such groves are called in the west, grew increasingly numerous, dotting the country to give it the semblance of one vast parkland. The weather held calm, cool, and crisp, and none of the dreaded dust storms we had heard so much about rose to bedevil us.

Late on a mid-September afternoon we came in sight of our new home port. Straddling the broad and muddy South Saskatchewan River, Saskatoon's church spires, grain elevators, and taller buildings loomed on the horizon like the masts and funnels of a distant fleet immobilized on a golden ocean.

Founded three decades earlier as a Methodist temperance colony, Saskatoon quickly outgrew its natal influences. By the time of our arrival, it had burgeoned into a city of thirty thousand people embracing the beliefs and customs of half the countries of the western world. Many of these, especially the Doukhobors, Mennonites, Galicians, and Hutterites, would prove to be mystery distilled in the eyes of a twelve-year-old from the staid Anglo-Saxon province of Ontario.

While Angus searched for a house to rent, we lived aboard *Rolling Home* in the municipal tourist park which

*The closely related trembling (or quaking) aspen and cottonwood poplars are the dominant trees in Saskatchewan northward to the edge of the boreal forest. Cottonwood is so-called because, in early spring, it covers the ground beneath it with tiny seeds attached to delicate parachutes of what looks like cotton wool.

was attached to the city's exhibition grounds and zoo. There were no other "tourists" so we had the place to ourselves except for two camels, an ancient buffalo bull, and a pair of elk who, together, made up the population of the zoo. I was delighted with these creatures whose like I had never seen before and was convinced by their presences that Saskatoon was truly the gateway to a wilderness world.

Angus was also pleased with his first taste of our new home, but Helen had her reservations. "The place all looked so *new* and, well, *temporary*, as if it had been thrown up yesterday and might all blow away tomorrow. All those boxy little bungalows covered with grey stucco and those great, wide streets with numbers instead of names, and the wind whistling down them from the North Pole in winter and up from the desert in the summer. It didn't look like Heaven to me."

It looked a good deal more like hell when, a week later, a dust storm struck. Soon after noon an oppressive darkness began to overshadow the city. Within an hour the sun had been obliterated and it seemed as if Saskatoon must suffer the same fate. Rising in the new deserts of the south-west and lifting high on the autumnal winds, the desiccated soil of the prairies drove over us. Helen lit the oil lamps in *Rolling Home* but so much dust was suspended in the air that the lamplight was diffused into an eerie glow. Everything we ate or drank that day, that night, and most of the next day tasted of the earth. It was an awesome experience.

Our rented house, when we finally found one, turned out to be one of the "boxy little bungalows." It was small and poky, and its only claim to distinction was that it belonged to a player from the Saskatoon Quakers, a hockey team of some eminence in western Canada.

As soon as we were more or less settled, I was sent off to Victoria School (the second of that name for me), and I seized upon this God-given opportunity to rid myself of my Christian name by registering as William McGill Mowat. My parents were somewhat bewildered when they discovered what I had done but permitted the sobriquet to stand. Billy Mowat I remained throughout most of my time in Saskatoon.

Apart from school I had little to do with my peers that first winter in the west. Most of the neighbourhood boys skated and played hockey at every opportunity. I did neither. Furthermore, I was not prepared either to learn to skate or swing a stick, since I knew I would certainly make a fool of myself if I tried. When pressed by the physical training teacher at Victoria to turn out for hockey, I asked to be excused on the grounds that I suffered from a condition of the inner ear which so disturbed my equilibrium that, if I moved rapidly or turned sharply, I was liable to faint and/or throw up. He believed me because, I suppose, it would never have occurred to him or anyone else in Saskatoon that there was a boy alive who wasn't mad crazy to become a hockey star.

Although my refusal to play hockey put me outside the pale, I did get to know one boy on our block. The relationship was short-lived. One Saturday afternoon down in our cellar, he introduced me to bestiality, onanism, and homosexuality all in one fell swoop by first masturbating his dog, then himself, and finally me. He was successful with himself and the dog but gave up on me and then delivered the shattering opinion that my dick wouldn't work because it was too small. The truth was that I was terrified of discovery, for my mother was in the kitchen overhead and might have descended the cellar stairs at any moment in order to attend to our fractious furnace.

Coal was the universal heating fuel in Saskatoon and was of an inferior variety mined mostly at Drumheller, Alberta. Keeping the home furnace going was a skill that had to be acquired by men, women, and older children as a matter of survival. The proper adjustment of the two control chains which led from a panel on the baseboard of one of the main-floor rooms down to the reluctant dragon in the cellar was a fine art. One chain controlled the draft on the furnace door and the other a damper in the pipe leading to the chimney. Both had to be perfectly adjusted, and both were skittishly responsive to a score of fluctuating factors including wind, outside-versus-inside temperature, the amount of ashes in the grate, and fuel in the fire box. Any one of these, if out of synchronization with the others, could upset the whole delicate balance and result in a dead fire and a frozen house. Keeping the dragon's maw stuffed with coal, and shovelling out and carrying away its ashy excretions could also only be neglected with dire consequences.

Especially when the outside temperature plummeted to 50° below zero Fahrenheit.*

---

*Despite the attempts of the Canadian government to force its citizens to switch from Fahrenheit to Celsius, I resolutely continue to use the Fahrenheit scale.

That first season was a truly chilling revelation of what a western winter could be and do. All through January and February, the thermometer remained below zero — usually twenty to forty degrees below! The cold seemed paralytic at first. It caused Helen intense pain by inflaming her chronic neuralgia whenever she ventured out. This was an affliction with which she had to bear throughout our years in Saskatoon.

Angus and I reacted to the ferocious cold with an excitement that escalated as the mercury dropped. We began romanticizing about being in the Arctic. He read all the books written by polar explorer Vilhjalmur Stefansson, while I buried myself in stories about the Hudson's Bay Company and the exploits of the intrepid Royal Canadian Mounted Police in the Far North. The high point of the winter for us was the day the thermometer fell to 52° F below zero.

Little or no wind accompanied these periods of excessive cold and the air became so still it seemed frozen into a crystalline solid, marbled by luminous blue columns of coal smoke and condensation slowly rising from heated buildings. Most automobiles would not function at such temperatures but horses could and did. By mid-winter Saskatoon had become a horse town and frozen horse balls abounded, to the great delight of boys who used them as pucks for street hockey and as projectiles against any suitable target, including the fancy fur hats affected by affluent businessmen.

The onset of spring was heralded not so much by the arrival of robins as by the reappearance of private automobiles from winter hibernation in garages and sheds where they had slept, jacked up on wooden blocks to preserve their tires. Eardlie was one of the first to emerge in 1934 and, seized by exploration fever, Angus set out in our little flivver to explore the world beyond the city. Mother and I went along on the initial trips but she opted out after our first encounter with gumbo.

The deep, rich prairie soil could and did (when conditions were right) produce bumper wheat crops, but when the frost came out it turned into a glutinous substance resembling the contents of the La Brea Tarpits. This was gumbo, and most Saskatchewan roads became gumbo quagmires every spring.

It took Angus a remarkably long time to realize that only horses and things with wings could deal with gumbo. On one occasion he got Eardlie so deeply mired it required four big percherons to extricate him, and they almost pulled the little car apart in the process.

Although I did not find many chums of my own species during that first winter, I was not without companionship. My parents got to know a professor of biology at the University of Saskatchewan who, recognizing my interest in animals, presented me with a pair of white rats.

I found these creatures so captivating that I gave them the freedom of my room. We happily shared the space until the day Helen discovered the female rat nursing eight or nine naked ratlets in a nest constructed inside one of my pillows.

The rats were banished to the cellar but my fascination with them continued unabated. In January of 1934 I wrote a letter to my paternal grandparents. "I suppose you have heard of my White Rat Company. It is progressing well and all orders for young ones will be accepted. I have never heard of such rapid breeders. Whew! I am told by a professor that if one pair of rats and their youngsters and theirs etc all breed, they will produce a total at the end of the year from 1500 to 2000."

Not everyone shared my high regard for the little rodents. Most of our neighbours believed all rats — white, black, or varicoloured — were vermin and ought to be exterminated. When rumours about our cellar tenants got around, threats were made to report us to "the Public Health." My father was annoyed at what he took to be an infringement on our privacy. I was indignant at what I regarded as rampant prejudice, if not racism, and began my first public crusade for animal rights. The following appeared in the "Victoria School Record" early in 1934.

If you were to ask me to name an interesting pet that can be kept in a small house I would immediately reply "the White Rat". This small animal has helped mankind more than we can guess. When Pasteur was attempting to find a cure for rabies the rat played perhaps the most important role of all.

It was this little creature that took the deadly injections of dried rabbit brains by which Pasteur was able to determine whether his cure was effective... Most hospitals now have a room set aside for breeding White Rats for medicine. They give their lives that ours might be saved and although you could hardly call them heroic, their great service to mankind will never be forgotten.

Not only are they useful but they are very amazing as pets.
They are exceedingly loving toward each other and when a male
and female are separated for a few days they show every possible
affection when re-united... Almost everybody who comes into
contact with White Rats in a very short time becomes keenly
interested and warmly affectionate toward them.

Billy Mowat

By March my rats had demonstrated their affection for
one another so successfully that the consequent population
explosion was making our basement smell like a barn.
When I could find no takers for the new generations (either
by sale or gift), Angus reluctantly lowered the boom and my
rat friends were exiled to the biology building at the university. I missed them very much for a time but, as spring
exploded, found new interests to distract me.

No house in Saskatoon was more than a few blocks distant from the open prairie. When I stepped off the sidewalk
at the end of 9th Street, I walked into a world not yet totally subjugated by Man and the Machine. Although most of
the aboriginal short-grass plains had been replaced by wheat
fields, enough remained, together with bluffs and sloughs,
to sustain an astonishing array of natural life.

This life reached a peak of abundance in spring. Ducks
and geese of a dozen species crowded the sloughs in such
numbers that the roar when a big flock took wing was like
the thunder of an express train. Mudbars on the river
became so packed with sandpipers and plovers en route to
their Arctic nesting grounds that, when they took off en

masse, it looked as if the bar itself was lifting into the air. The poplar bluffs, redolent with the scent of balsam from sticky buds, and frothed with the transparent green of budding leaves, were metropolises of bird and animal life. Crows, magpies, hawks, and owls busied themselves building their nests amongst the stouter branches of the aspens while kaleidoscopic mobs of migrating warblers scoured the limbs and twigs for insects. Blackbirds, shrikes, and catbirds contested for nesting territories in the wolf willows surrounding the bluffs, and meadow voles, thirteen-striped gophers, and Franklin's ground squirrels rustled and whuffled through the cottonwood "snow" which lightly coated the floor in each of these separate little forests.

The open fields, whether composed of last year's stubble, ploughed land, or summer fallow, seemed equally alive with common gophers,* garter snakes, meadowlarks, pipits, grey partridge, and sharp-tailed grouse. Insects were also having their heyday and prairie crocuses bloomed everywhere.

This was a world so thoroughly and vibrantly alive that I could hardly have avoided becoming enamoured of it even had I entered into it as a total ignoramus. As it, happened, I found a guide who introduced me to many of its marvels.

One Saturday late in May, I emerged from a bluff to be confronted by a tall, unshaven man with a hawkish face. He was wearing a brilliant crimson shirt and loudly singing the

---

Officially: Richardson's ground squirrel.

refrain from a Scots musical comedy while briskly keeping time with a huge butterfly net.

I could not have been more surprised. In my small experience, "bug catchers," because they were generally mocked by right-thinking people, were unobtrusive and self-effacing to the vanishing point. Yet here was one who strode across the land as if he were lord of all he surveyed. He stopped abruptly and stood arms akimbo while he took my measure. He noted the shiny brass tubes of the venerable (Boer War vintage) field glasses that hung by a cord around my neck, then he introduced himself.

"Alisdair McPherson at ye'r service, laddie! But ye may call me Tom. Tell me now, what rare creatures have ye spied the day?"

Tom's father had been a ghillie on one of the vast highland shooting estates, making his living guiding the gentry in their slaughter of red deer and grouse. But Tom and the gentry had not taken to each other so he had come out to Canada, where he served an apprenticeship as a baker in Toronto, before getting married and travelling on to Saskatoon in pursuit of the chimera of "the wide open spaces." Barely a year after I met him, he and his little family moved on but in the meantime I was his grateful acolyte.

He was a self-taught naturalist who collected butterflies, moths, and other insects, not to add to the repositories of scientific minutiae but to gain insight into their lives. He did not know their Latin names but he had an amazing understanding of their ways and habits. It was the same with birds. Although he collected birds' eggs ("Ye may take one of each kind, mark ye laddie, and nae mair!"), it was the living birds that fascinated him and he knew them with an intimacy that stemmed from empathy and not from books.

Tom worked the night shift at the biggest bakery in town and spent most of the daylight hours roaming the prairies in any kind of weather. I don't know when he slept or what sort of a family life he had but I know where his primary allegiance lay. It was with the world of the Others, and he took me with him into that world.

———————————

By early June the few showers that had enlivened the spring had ended. Day after day the pallid skies remained unmarred by the slightest trace of cloud. The West was facing yet another year of drought and, although Saskatoon was somewhat to the north of the Dust Bowl (as the devastated southern plains were now being called), the prospects for the months ahead looked grimly arid.

What people wanted — what they craved — was water. Not just drinking and washing water which, God knows, was scarce enough, but *visual* water. The very sight of a significant body of water, even if it was no more than a slough filled with an alkaline slurry so bitter it made one gag, brought solace to the spirit. People dreamed they were dying in a desert and, when they woke to the burning desiccation of a Saskatchewan summer day, yearned for the sight of water with such passion that they almost became unhinged. Everyone was an aquaphile and those who had it in their power to do so fled the city for whatever body of water they could find. Having reached it, they would build or rent shacks or shanties which, for the most part, were mere shelters from the baleful glare of the sun.

Even before school ended we had joined this weekend exodus. Because we owned a *mobile* shanty, we were able to visit and assess all the available watering holes. Most of these lost their allure on close acquaintance. We tried Pike Lake and found it to be a glorified slough so weedy that only a pike could have lived in it. Watrous turned out to be an even bigger slough whose water was so salty — *Epsom* salts, mind — that one actually could not sink in it. Wakaw was tolerable, except that swimming in it gave us scabies. Jackfish Lake was so shallow one could walk three-quarters of the way across it, and it felt and looked like warm soup.

We three were not unaccompanied on these excursions. Shortly after my thirteenth birthday we had acquired a dog. His name was Mutt. Since I have already written an entire book about him, it will here suffice to say that he became an integral if not dominant member of the family. However, getting to know central Saskatchewan was only part of the experience of that year. Getting to know Mutt was an even more important part. Until his death, he was to loom as large in my life as any other being ever did.

NINE

Far from being satiated by the voyage to Saskatoon, Angus's appetite for distant venturing seemed only to have been whetted. During our first winter in the west, he began putting things in train for an even more adventurous expedition.

How he managed it I do not know, but he was one of the most persuasive human beings ever born, and after only a few months at his new job he managed to con the sage and elder members of the library board into granting him six weeks' holiday with pay, the same to begin in mid-July. Angus intended to use this time to make a pilgrimage to the largest body of water in the world — the Pacific Ocean.

Originally he had planned to make the voyage in *Rolling Home* but a middle-aged bachelor named Don Chisholm, with whom we had become friends, talked him out of it. Chisholm was a divisional superintendent with

Canadian National Railways and had once supervised the section of the main line which crawls precariously over the Rocky Mountains.

"Angoos," he said — his accent was still strong — "ye'll no haul yon caboose out o' the foothills. But think, mon, if ye did manage to drag it awa' up into they mountains ye'd likely end up, the lot of ye, tipped into the bottom of aye canyon a thousand feet doon, sprouting wings afore ye'r time."

Angus was not noted for heeding advice but this time he did so. We left the Ark behind.

In consequence Eardlie was appallingly burdened. None of your glass-and-chrome showcases of today could have carried that load a single mile. The cargo included a large umbrella tent and three folding wooden cots tied to the front mudguards. A huge wooden box was fitted on each running board (the running board was an invaluable invention long since sacrificed to the obesity of modern cars) and filled with, amongst other things, pots and pans, suitcases, blankets, and a gunny sack containing shreds of cloth which, in her spare time, Helen worked into hooked rugs.

The roads were tunnels of dust as we drove south and west, and it became so thick that we three human travellers had to wear motorcyclist's goggles. One evening Angus decided this was favouritism and Mutt should have the same protection. We were then entering the outskirts of Elbow, a typical prairie village with an unpaved main street almost as wide as the average Ontario farm, and two rows of plank-fronted buildings facing one another distantly across an arid emptiness.

Angus, Mutt, and I entered one of the shops together and when an aged clerk finally appeared from the back premises, Angus asked him for driving goggles. The old fellow searched for a long time and finally brought out a pair which had been manufactured during the first years of the automobile era. They seemed serviceable, so without more ado Angus tried them on Mutt.

Happening to glance up while this was going on, I caught the clerk's gaze. He was transfixed. His leathery face had sagged like a wet chamois cloth and his tobacco-stained stubs seemed ready to fall from his lower jaw. Angus missed this preliminary display but was treated to an even better show a moment later as he got briskly to his feet holding the goggles.

"These will do. How much are they?" And then, remembering he had forgotten to pack his shaving kit, he added, "We'll need a shaving brush, soap, and a safety razor too."

The old man had retreated behind his counter where he pawed the air with an emaciated hand for several seconds before replying.

"Oh, Gawd!" he wailed, and it was a real prayer. "Don't you tell me that dawg *shaves*, too!"

Our second night on the road was spent at Swift Current, well into the Dust Bowl country. The place had a lean and hungry look. We were very hot, very tired, and very dusty as we drove through it seeking the municipal tourist camp which, we hoped, would provide running water. Since motels and tourist cabins did not yet exist, there was no alternative to camping out unless one rented a tiny cubicle in one of the wooden crematoria that passed as hotels in these small towns.

Swift Current was proud of its tourist camp, which was located in a brave but pathetic attempt at a park beside an artificial slough. We contemplated going for a swim but changed our minds when we discovered a dead dog — a *very* dead dog — floating amongst the weeds.

Only a drizzle of tepid water could be tempted from the single faucet which was intended to supply the needs of tourists, so we went to bed unbathed and uncooled. We had to keep the tent buttoned up tight because the mosquitoes rising from Dead Dog Slough were numerous and hungry enough to have drained the blood from a cow. Mutt refused to sleep outside so the four of us tossed and muttered all night long in mutual misery. Camping on the prairies that summer was not a thing of joy forever.

The remainder of our journey across the scorched plains was a monotony of mounting fatigue and tempers strained by the long pall of dust hanging over a yellowing wasteland of dying wheat. Poplar bluffs were few and far between and their parched leaves rustled stiffly as if already dead. The sloughs were almost totally dry. Here and there a puddle of muck still lingered but these pot-holes had become death traps for innumerable ducks because botulism throve in the stagnant slime. The ducks died in their thousands and their bodies did not rot but withered as mummies do. It was a grim passage and Angus drove Eardlie hard, heedless of the little car's boiling radiator and labouring engine.

Then one morning brought a change. The sky that had been dust-hazed for so long grew clear, and ahead of us we saw the violet shadows of the Rocky Mountains hanging between land and air.

We camped early that night and were in high spirits at the promise of escape from drought and desert. When the little gasoline stove had hissed into life and Helen was busy preparing supper, Mutt and I went off to explore this new land. Magpies rose ahead of us, their long tails iridescent in the light of the setting sun. Pipits climbed towards the high, white clouds and sang their intense little songs. Prairie chickens rocketed, chuckling, from behind a trim farm-house. We walked through a woods whose leaves flickered and whispered as live leaves should.

By evening of the next day we were climbing into the foothills. The ground seemed to be swelling under Eardlie's singing tires like great oceanic combers lifting us towards a distant coast of soaring peaks. This was a new kind of prairie: undulating, cut with valleys and ravines, green and alive, and dotted with small herds of cattle. It lifted all our hearts with expectation of what might lie ahead.

In the last week of July we began the passage of the mountains, having chosen the northern route which, at that time, offered no easy path even for a Model A. The roads were narrow, precipitous, and surfaced with loose gravel. Often there were no guardrails and we would periodically find ourselves staring over the naked edge of a great gorge while Eardlie's wheels kicked stones into an echoing abyss.

The mountains frightened me. I felt dwarfed to the vanishing point and hypnotized by the sheer bulk and majesty of these mighty eruptions of the flat earth I had previously known.

I think all four of us were a bit on edge, anticipating the possibility of some horrendous happenstance. And something horrible did happen. Not long after we had begun a cautious descent of the Selkirk Range on a road that was hardly more than a narrow shelf cut into the side of a mountain, we were overtaken and closely passed at great speed by a big touring car with its top down. Gaily dressed young men and women waved bottles at us, laughing and shrieking with delight as they plunged down the switch-back road. Shaken, Angus cursed them and braked to a halt on the narrow verge, but my eyes were glued to the careening car ahead. There was a blur of movement, as of a bird taking wing, and the big car vanished.

I don't know what got into me. I may have thought that somehow I could help. I jumped out of the rumble seat and set off down the steep gradient at a run. Angus and Helen yelled at me to stop, to come back, but I ran on with Mutt pounding along beside me. As I neared the point where the car had disappeared I began to lose heart. I went the last little way in slow motion. And then I was looking over the edge.

The car had landed upside down on a ledge of rock a hundred feet below the road. Its wheels were still turning. A little coil of smoke was rising from it. But there was not a sound! Not a whisper of a human voice...nothing but the purl of an unseen stream in the distant bottom of the gorge.

By the time Angus and Helen reached me I was being sick to my stomach. Then a truck full of wardens from a nearby national park pulled to a halt. Men piled out, stared into the canyon for a moment, then some of them began trying to find a way down. Their leader came over to us. He spoke kindly.

"Mister, you'd better take the kid outta here. It ain't goin' to be a pretty sight. When you get down to Revelstoke tell the police whatever you saw happen."

What we had seen happen, as we were to discover later, was seven people reduced to mush in the space of a few seconds. This dreadful incident, echoing as it did Don Chisholm's dire warning, increased my fear of the great mountains almost to the point of terror. I was glad when we left the shadows of the brooding peaks to descend into the Okanagan Valley where we hoped (and I expected) to see the Ogo Pogo, the legendary monster of Lake Okanagan. Unfortunately, the dragon proved reluctant. We solaced ourselves by gorging on the magnificent fruits for which the valley is also famous.

To our surprise, Mutt shared our appetite and for three days ate nothing but fruit. He preferred peaches, muskmelon, and cherries. Cherries were his undoubted favourites. At first, he had trouble with the pits but soon perfected a rather disquieting ability to squirt them out between his front teeth. Not everyone appreciated this. I recall the baleful look directed at him by a woman passenger on the little ferry in which we crossed the Okanagan River. Mutt was certainly a vision to give one pause as he sat in the rumble seat, goggles pushed high up on his forehead, eating cherries out of a six-quart basket. After each cherry he would raise his muzzle, point it overside, and nonchalantly spit the pit into the green torrent of the river.

My most cherished images of the mountain country were not of the forbidding peaks but of the animals. There was very little human traffic on the recently constructed roads but the Others had been quick to make use of these new pathways. Several times we had to yield the right of way to herds of elk; twice to parties of mountain goats (who stared at us with a haughty golden glare); and once to a grizzly sow with her offspring — she rose up on her hind legs in the middle of the road and looked *down* upon us as we crouched apologetically in our little car. In addition, we saw scores of black bears; a lithe shadow slinking into a stand of lodgepole pines — I was convinced it was a cougar; my first-ever wolves (a family seen distantly as they crossed an alpine meadow in single file); and innumerable smaller creatures including rabbit-like rock conies, chipmunks, and squirrels of kinds I had not met before.

Then there were the birds. That spring Angus had bought me a field guide and with its help I identified forty species that were new to me, ranging from the minute miracle of a rufous hummingbird to golden eagles soaring both below and above us as Eardlie nervously negotiated cliff-clinging roads. I did not know how extraordinarily lucky I was to be seeing the living world of the high rock country in an abundance of numbers and variety of kinds almost as great as had existed before the arrival of European man. Happily, I could not know how terribly they would be diminished within the span of my own lifetime.

In due course we reached Vancouver and saw the sea, but did not *meet* it until we came to rest on a strip of beach on the coast of Vancouver Island.

We camped on Rathtrevor Beach for three weeks and I discovered another new world. This one was dominated not by mountains but by the mysterious comings and goings of the tides; of huge translucent swells bursting upon offshore reefs; of flotsam and jetsam ranging from the trunks of ancient trees ten feet in diameter to baseball-sized, green glass Japanese fishing floats which had drifted across the almost unimaginable width of the Pacific.

In short, I entered into the world of ocean. It and its denizens absorbed me so completely it is a wonder I did not begin to grow gills. At low tide I burrowed in sand and mud flats in pursuit of marine worms, clams, and a hundred other creatures of bizarre form and colour. At high tide I was a beachcomber pausing to stare seaward through my field glasses at the flashing fins of what were probably killer whales, and at the bewhiskered faces of seals, who stared right back. Sea birds abounded, and so did land birds in the lush rainforest behind the beach, which seemed tropical to me. I was in seventh heaven. As Helen recalled my state of being:

"We almost never saw you, and I'm sure *you* never saw *us* at all. Half-naked or wholly so, you scampered about like a water sprite, paying no attention to anything except your

precious crabs and shells and birds. We thought we might
have to tie you up to make you leave."

Indeed, leaving Rathtrevor and the ocean was an enormous wrench. I ought to have been born beside the sea —
any sea — and I have a strong suspicion that this inclination is
present in all human beings. The ocean calls us for it is the
ancient and original home, not only of mankind but of all life.

When the dread time came for our departure, I determined
to take as much of the sea world with me as I could. My
booty included clams, mussels, giant whelks, and shellfish of
all kinds; starfish; crabs; anemones; sponges; even seaweeds.
Some of these I packed into a carton which was stowed
under my feet in the rumble seat, but many others I hid
away in nooks and crannies of Eardlie's interior to avoid
Angus's dictum that I could add only a token amount to our
already heavily overladen car.

Many of the poor creatures I collected had been alive
when removed from their home. They did not long survive
the heat and desiccation of our homeward journey. Shortly
they began to stink. At first this amounted to no more than
an occasional whiff of decay which I, with admirable presence of mind, blamed on Mutt: "He's been rolling in something." This was a credible explanation for a time because
Mutt loved to perfume himself wih substances offensive to
the human nose. However the truth came out when a
thunderstorm forced Angus to put up the seldom-used canvas roof over the front seat. In the process, he unfurled a

clutch of deliquescent starfish, one of which slithered into my mother's lap.

My parents were good about it. Instead of ordering me to jettison my precious specimens, Angus procured an orange crate in which most of my defunct sea creatures were interred, well wrapped in newspaper. The crate itself was then lashed to the rear bumper. At each day's end, it was my task to carry the crate a suitable distance away from our campsite and cache it, to be retrieved only shortly before our departure the following morning.

By the time we made the fifth camp of our homeward journey, not far from Banff, the crate had developed a life of its own and was, to say the least, redolent. As soon as we had parked I dragged it off to a clump of trees, keeping one hand free with which to hold my nose.

That night the inevitable happened. We were awakened by the most frightful uproar. It seemed to consist of the furious barking of a dozen dogs, mingled with the roaring of a menagerie of lions. It was, in fact, poor Mutt trying desperately to defend my specimens against a couple of black bears whose investigative interest was not to be denied.

This was not the time for heroics. We collared Mutt and abandoned camp, leaving the tent and most of our belongings behind. Then we drove to the nearest warden's cabin in search of reinforcements.

The warden did not receive us sympathetically.

"Putting that stuff out where the bears could get to it was a damned fool thing to do," he told us sternly. "Now like as not, they'll have a taste for seafood and who in hell can tell where that will lead?"

We had no answer, but I had a quick mental image of a pack of irate bears showing up in Vancouver looking for a seafood restaurant. And it would be my fault.

TEN

We arrived back in Saskatoon in late August and moved into a new home a week later. This one belonged to a Professor Morton who had gone on a sabbatical after, perhaps foolishly, renting his house to us.

It stood on élite Saskatchewan Crescent, a street of many mansions high above the broad river valley, commanding a sweeping westward overview of the more plebeian reaches of the city. The property boasted a big back yard containing a garage and an ornate gazebo, both surrounded by a high wooden fence. Impeccably finished inside and out, our new home was a far cry from the hockey player's grungy little box.

"Tony!" was the admiring epithet Grandmother Georgina Thomson bestowed upon the place when she and Hal arrived by train for a visit later that autumn. For me the house's greatest virtue lay in its proximity to the steeply wooded river bank, one of the few places inside the city limits to have escaped significant human "improvement." It was a ribbon of natural parkland — a birch, poplar, berry bush, and bramble jungle a quarter of a mile wide, that seemed to stretch to infinity in either direction along the river valley. Although no longer home to bison or to the great grey prairie wolves, foxes still frequented it as did the occasional coyote. And as did I on every possible occasion, for this was wilderness at my own doorstep.

Late in September my parents hired a maid. Rachel was nineteen, only six years my senior. Slim and quick, she was gifted with the dark vivacity of the northern Celts. I immediately fell in love with her, not in any overt sexual way but as with the sister I had never had. And I think Helen came to feel Rachel was the daughter she did not have. In any event, the newcomer soon seemed as much a part of our small family as if she had been born into it.

Half a century later, she wrote to me the first word I had heard from or of her in all that time since.

"Sitting here with pen in hand going back over the time I was with you and your family it is hard not to include my whole life as it was then. My Dad was a wheat farmer, and a good one, though he had come from Scotland as a middle-aged man and had to learn everything about life on the prairie.

"I was born and grew up on the farm and never knew what it was not to have all I wanted or needed as a teenager in the late '20s — the fun days of the Flapper era. Then in 1929 the Crash came and soon we were well into the Dirty Thirties with the drought and dust and no price for what wheat we could grow. My Dad barely made enough to pay the men who worked for us, and keep us all from going on relief.

"We had to scratch for a living then but there was almost no paid work available. At fifteen, I had to stop school and help at home until things got so bad there was not enough to eat for all of us. So I went off to Saskatoon and did housework. I started at $25 a month but as time went on and things got worse and worse, the going wage by the time I knew you was down to $10 a month.

"I took night courses in Saskatoon and got my certificates in shorthand and typing but it was no use. The few secretary jobs that came up all went to young men. I was lucky to have $10 a month and my keep. Many of the girls I grew up with did not even have that much and some had nothing at all.

"From age seventeen I had been engaged to be married but my young man could not get work either. We waited until 1933 then got married anyway. We decided neither one of us could be any worse off than we were, so no matter what we did it had to be better. But we could not afford to live together, of course. He went away to The Pas in northern Manitoba as a logger in a bush camp and I did housework in Saskatoon but at least we *felt* married and we were together in heart and spirit.

"That was when I came to you people. You were a nice easy family and there are lots and lots of pleasant memories from that time.

"Coming from Ontario, you had to adjust to a new school and playmates and didn't do very well at that. Your dog was your best friend. Saturday mornings you and Mutt would disappear down the riverbank into the bush, or out on the prairie. Sometimes you'd come back with your pockets full of owl pellets, balls of bones and feathers. And you'd expect me to be interested in your finds. Well, I could relate to what interested you. I had roamed our farm all my childhood and knew every kind of bird and where their nests were, and had followed their summers from eggs to fledglings.

"School wasn't something you were interested in, so we spent many evenings with me trying to teach you what you couldn't be bothered with at school. That wasn't all of your troubles. You remember the Catholic school nearby? The boys from that school used to often chase you home. You always played innocent but I used to wonder what you did to get them so mad that half-a-dozen of them on their bikes would come flying after you right into our front yard.

"Your mother was in bed much of that winter with infections of her sinuses, so I more or less ran the house. She gave me $40 a month for groceries and other supplies like coal that were delivered mostly by horses. She also gave me the Boston Cook Book she had when she got married, and you and I learned to be pretty good cooks.

"Your dad was taking a night course at the university so you and I spent lots of time together. I did your maths for you and you taught me a lot I didn't know about birds and animals; but sometimes we would just sit and talk. I guess we were both lonely. You used to choose books for me to read, and some of them were not exactly what my Dad would have approved of.

"I remember the night I was going to visit an old Scots couple on the other side of the river and you came along. It was early winter and the river was already frozen. You decided we should walk across the river instead of going the long way round by the bridge. It would save streetcar fare, and I couldn't afford that anyway so I agreed. It was very dark and about twenty below zero and we kind of felt our way across. When we got to the old folks' house the smell of mash that came out the door could have knocked you down. I knew what it was but old Charley didn't want you to know so he said it was grain soaking for his chickens. And you never even smiled as you said, 'I hope it doesn't make them too drunk to walk.' Charley gave us what-for for crossing the ice. The day before, it had smashed under a cutter and drowned the horse. We could have vanished without a trace and nobody the wiser.

"That was the Xmas you woke me at 4 in the morning, crawling into bed with me to show me the little .22 your father had bought you. What a way to be wakened up, with an icy cold rifle to be admired!

"The last time I saw you was in the fall of 1935. I met you in the street with Mutt. You asked me if I was coming again to spend the winter with you and when I said I couldn't, that I was going to join my husband in The Pas, you looked sad and asked, 'Who's going to help me with my math, and who can I talk to at night?' I guess you were still pretty much of a loner."

Indeed, I was something of a loner but by no means a misanthrope. It was just that I could not find many people my own age who interested me or shared my interests. Those I did find tended, like the Marsh Boy, to be unusual.

One such was Fred. His father was a locksmith who spent what little he earned on booze for himself and his wife, and used to beat his son on any or no provocation. I first met Fred one autumn Saturday when he was snaring gophers in a stubble field north of the city. I saw him before he saw me and, being a prudent soul, stopped to study him from a safe distance through my old field glasses. He was not a reassuring sight. He was clad in tattered bib overalls, had tangled hair hanging to his shoulders, and a truly fearsome face which, I later learned, had been disfigured at birth. I would have made a strategic retreat had he not looked up, seen me, and waved a friendly arm. Somewhat apprehensively, I approached.

"Hi ya, kid! Wanna make some money? Farmer owns this half section is gonna pay a cent a tail for all the gophers I can catch. There's plenty for both of us, an' I got lots of binder twine to snare 'em with."

This was a truly generous offer and I did not refuse. Fred and I became good friends. He did not have many such. He was called Frankenstein at school, where he was generally treated with aversion, yet he was a kindly and harmless boy, avid for whatever I could teach him about prairie creatures. In return he taught me how to open locks and gave me a beautifully made set of picklocks with which

I could have set myself up as a housebreaker if I had had the urge or, as it may be, the courage.

Fred introduced me to a girl of our age who was almost an albino and so was nicknamed Whitey. Whitey's mother was a single parent. How she made a living was no concern of ours, but we knew how Whitey earned *her* disposable income. For five cents, or a chocolate bar, she would accompany a boy down into the jungle of the river bank and show him her "private parts." For five cents *and* a chocolate bar, she would allow the youth to attempt a fumbling kind of coitus — standing up. Whitey had her standards. She would not lie down with a boy, claiming indignantly that this was "dirty stuff." Whether she meant this literally or figuratively I never knew, but I do know that her favourite brand of chocolate bar was Sweet Marie.

My last meeting with Fred was a shattering experience. In the summer of 1935 a circus came to town. My father took me to it and I insisted on seeing the freaks on display in the midway. These included the inevitable fat woman, a three-legged goat, the World's Skinniest Man — and the Wild Boy from Borneo.

The Wild Boy was a fearsome sight. Clad only in a breech clout of mangy fur, he squatted in a cardboard cave, growling ferociously and occasionally gnawing scraps of raw meat from a bone. His body was streaked with red and purple paint and dirt. His hideously contorted face was also painted. I recognized him anyway.

I grabbed my father's arm. "That's Fred!" I said with horror in my voice.

"Can't be," my father replied soothingly. "This chap's from Borneo."

But I knew. I leaned forward over the low barrier in front of the cave and whispered nervously, "Fred! What are you *doing?*"

He looked at me, growled menacingly, then muttered just loudly enough for me alone to hear.

"Makin' a dime, Billy. Makin' a dime."

Presumably Fred left town with the circus because I never saw him again. Well, life with the circus crowd could hardly have been worse than what he had had to endure at home.

———————————

About this time my parents began to worry that I was not (as we would say these days) becoming socially integrated. They set out to remedy this by making me attend Sunday School at the Anglican Cathedral where I would be associating with "lots of nice children" of my own age and kind. The cathedral was on the other side of the river, a long bike ride and an even longer walk from home. I went reluctantly (pocketing the ten cents given to me for the collection) until cold weather began, after which I found a sufficiency of excuses to stay home. "Alibi Ike," my father called me with some admiration, for he was secretly on my side, never having been able to abide Sunday School himself.

It was then decided I should join the Cubs. I did not like this either. I had read and re-read Kipling's *The Jungle Book* (upon which Baden-Powell based much of scouting's ritual) and for me Akela and the Wolf Pack were jungle-dwelling realities — not prosaic human imitations thereof. I think I went to three meetings, during each of which we Cubs had to squat in a circle around a distinctly obese (and un-Akela-like) Cub Master, and engage in a group cacophony which was supposed to represent the howling of a wolf pack. I found the procedure embarrassing if not downright silly.

Saskatoon kids knew a lot of jokes about Cub and Scout Masters, and one evening at dinner I casually remarked that mine was "really chummy. I guess he thinks a lot of me because he keeps patting me on the behind." I was allowed to withdraw from scouting with no further ado.

The next attempt to socialize me saw me volunteered to Saskatoon's Little Theatre group as a bit player in a pantomime based on *Alice in Wonderland*. This was slightly more to my liking. Rehearsals were held in the spooky attic of a downtown warehouse. It had a complement of bats which used to swoop about when disturbed by our Thespian activities. The bats distressed the women and girls but delighted me, for there was a den of the little creatures in the attic above my bedroom at home and I was becoming something of a bat aficionado.

My role was not demanding. I played the Dodo, and all I had to do was shuffle across the stage, identify my character by name, then wander off into the wings. This verbal identification was deemed necessary because my costume, which was composed of angular masses of cardboard with chicken feathers glued all over them in random clusters, did not resemble any creature that had previously been seen on earth. The costume was also very difficult to manoeuvre, and was so constructed as to render me nearly blind.

On opening night I duly shuffled onto a brilliantly lighted stage, peered myopically through the inadequate eye slits, and set course for the opposite wing.

I didn't make it.

Instead I made sudden and forceful contact with the Mad Hatter, played by the male lead who was a policeman in ordinary life. I knocked him down and myself collapsed in a heap of crushed cardboard and a cloud of feathers. Then I belatedly identified myself.

"Dodo," I said as firmly as I could.

"DUMBO!" corrected the enraged Hatter with a roar.

At the after-theatre party my mother apologized to the cast on my behalf.

"He can't really help himself," she explained. "He gets it from his father. Has to be the centre of attention, no matter what. Some day they'll both fall through their pants and hang themselves."

Although I don't know exactly what it means, the shadow of this dire prophecy haunts me still.

My mother's final attempt to socialize me was to enroll me in the Vienna Children's Choir, a group of about thirty mostly female youngsters directed by an immigrant Austrian tyrant who may well have been a relative of Hitler's. I hated him and I hated the choir.

Rehearsals were held on Sunday afternoons and, since I could not be trusted to attend them on my own, Helen would personally deliver me to the auditorium. I pleaded to be allowed to stay home from the fourth rehearsal on the grounds that I *was* sick. Helen gave me a look of practised scepticism and dragged me off to catch the streetcar. But I really was sick. In the middle of "Ave Maria," I fainted and fell down, taking two other youngsters with me as I clasped them in a drowning man's grasp. That would have been disruption enough but my body overdid things. My bladder, which had been at maximum distension, emptied itself all over me and my two neighbours.

This may well have been the darkest moment in my entire life. Taken home in a taxi, I was put to bed with a diagnosis of influenza. I probably would have tried to stay in bed for ever had it not been for the happy accident that nobody else from Victoria School was in the choir. Even so, I was deadly loath to go back to school to face what could have been the ultimate ignominy. Fortunately, my luck held and the story of how I sabotaged "Ave Maria" never reached the ears of my peers.

At this juncture my parents gave up their attempts to find a suitable niche for me in Saskatoon's social structure. This was not entirely due to the defeats we had mutually endured. It was as much due to the fact that, having failed to find a tribe which suited me, I created one of my own.

During the autumn of 1934 I had begun to acquire a bit of a reputation at school as an interesting eccentric. Although many of my contemporaries (mostly in the hockey and baseball crowd) continued to deride me as a "sissy nature kid," others were impressed by the remarkable relationship which existed between me and Mutt, one which seemed to verge on the paranormal. It was claimed by some (and who was I to deny it?) that I could communicate with beasts by means of mental telepathy, something which was all the rage in those times. My possession of a witch doctor's bundle composed of dried tarantulas and poisonous centipedes (a present from my uncle Jack who had picked it up in Africa during the war) did my reputation no harm, and the fact that I was known to share my bedroom with bats was thought fascinating by some, if in a repulsive sort of way. These things produced a sufficient measure of regard from a small group of other youngsters to enable me to create the Beaver Club of Amateur Naturalists.

Initially the club consisted of four boys and three girls, one of the latter being Tom McPherson's daughter, Kathleen. All candidates for membership (excepting Kathleen, with whom I was enamoured and who therefore had dispensation) were required to submit to a rigorous initiation. Each had to be able to list from memory one hundred birds, twenty-five mammals, and fifty fish, reptiles, or insects. Each had to undertake at least one ten-mile nature hike a month. Each had to write, then read aloud at one of our regular weekly meetings, a four-page essay about nature. Finally, each had to donate a natural object of some considerable value to the Saskatchewan National Animal Museum.

I had actually begun the museum some time before founding the club and shortly after my first exploration of the professor's house. In the basement I had found a large, wood-panelled room whose walls were lined, floor to ceiling, with glass-fronted book shelves. Recognizing their potential as display cases for my ever-growing collection of bits and pieces of animate creation, I staked my claim on the room (which had been the professor's study) as a place in which to do my homework. Since neither Angus nor Helen had any other use in mind for it, they let me have it.

One of the first things the Beaver Club did was to devote a Saturday afternoon, when my parents were absent from home, to pulling the professor's hundreds of academic tomes off their shelves and hauling them out to the garage. We stacked them between the lawn mower and the lawn roller where nobody but mice was likely to discover them until spring. Then we began filling the bookcases with our exhibits.

Kids love collecting stuff, and enthusiasm amongst the club members reached fever pitch. During two successive weekends we laboriously hauled home scores of clay-coated bones from a landslip a mile away on the river bank. I identified these as dinosaur bones. They may actually have belonged to buffalo, but more probably cows. We also scoured barns, back sheds, and attics in the neighbourhood, from which we abstracted some rare and wonderful objects.

Until a few years ago I still had the "acquisition catalogue" in which was carefully inscribed a list of our exhibits. It has now gone from me but I remember some of the more interesting entries. There was, for instance, the joined skulls of a two-headed calf. There was an enormous umbrella stand made from the lower leg and foot of an elephant. And there was a gruesomely discoloured human kidney in a jar of alcohol, "borrowed" from his doctor father by one of our members.

We also had a stand of mounted tropical birds which had originally been protected by a glass bell jar. The jar had since been broken and the birds had become home to legions of moths and a myriad of skin-bone-feather-eating little beetles called dermestids. The display had, in fact, become a sort of mini-menagerie whose enterprising members quickly colonized the professor's house.

Another prize possession was a decrepit stuffed black bear cub which had languished for too long in a damp basement. It stank with a peculiarly penetrating pungency.

It was our intention to hold a grand opening of the museum during the Christmas holidays. We planned to invite the mayor and other dignitaries. I had even written an announcement to be sent to the *Star Phoenix*, in which I extolled the uniqueness of our endeavour: "It is the most stupendous collection of natural and unnatural curiosities of all sexes ever gathered in Saskatoon."

Disaster struck before we could go public. The moths and beetles precipitated things by invading the cupboard where Helen kept her raccoon coat and establishing a lively colony there. And a week of wet weather brought the bear to such a peak of pungency that my parents finally got wind of what was happening below decks. Angus made an investigation which was immediately followed by an ultimatum. We were given twenty-four hours to disperse the

Saskatchewan National Animal Museum or it would end up in the rubbish bin behind our house.

I was not entirely devastated by the loss of my museum. The collecting phase had come to an end in any case, and my tribe was losing interest. Moreover, I was already preparing to engage my little band in another enterprise.

In December of 1934 a new magazine joined the ranks of Canadian periodicals. As is the Canadian way, it did so without fanfare. *Nature Lore — The Official Organ of the Beaver Club of Amateur Naturalists* came quietly upon the scene. The cover portrayed a bloated sea-gull about to bomb a very shaggy beaver. The club's motto was inscribed below the beaver. *Natura Omnia Vincit* which, if my Latin is to be trusted, means Nature Conquers All.

The first issue contained: "*Stories and Articles and Poems About Animals, Birds and Reptiles, Together with Various Illustrations and Anecdotes by Members of the Club. Price 5 or 10 cents the Copy.*"

The variable asking price was, I submit, an act of genius. Those were Depression times so we dared not charge more than a nickel but, by giving the purchasers the *option* of paying ten cents in a good cause, we shamed most of them into doing just that.

It will come as no surprise to learn that Billy Mowat was the editor-in-chief. He wrote an impassioned editorial for the first issue, from which I quote:

"Birds and animals do not get heard enough in this country and are not treated well. The Beaver Club intends to do something about this. Every 5 cent bit contributed to this magazine will be spent on the betterment of the birds and animals of Saskatchewan..."

I also wrote most of the text, although I attributed many of the pieces to my loyal tribesfolk. It was the least I could do. They fanned out all over Saskatoon in fine weather and foul, hawking copies of the magazine to all and sundry.

The public's reception astounded us. Within a week the entire first issue, amounting to fifty copies, had sold out. The three subsequent issues, with press runs of a hundred copies each, did almost as well, earning a grand total of $25.45, which was more than many people were then being paid for a week's labour. This sum was almost pure profit because I had persuaded Angus to print *Nature Lore* on the library's mimeograph machine (using library ink and paper) as a charitable contribution to a worthy cause.

I believe it *was* worthy. The articles may have been a little didactic: "Planestrius migratorius [the robin] is a prominent local insectivor"; or somewhat overblown: "The crow can talk extremely well and is as intelligent as most people." Nevertheless, the effect was to at least engender some interest in and sympathy for wild creatures amongst people who had never previously given a thought to the possibility that they might have something in common with other animals.

As promised, we spent our money on good works. Each winter the discharge of hot water from the city's coal-fired electric generating plant maintained an open pond in the otherwise frozen river. This provided a haven for ducks and geese which were unable to join in the great south-bound migration because of sickness or, more usually, because they had been wounded by hunters. In previous years, most of these unfortunates starved to death long before spring came, but during the winter of 1934–35 the Beaver Club saw to it that they were well-supplied with grain and corn. In so doing we set a precedent which, I understand, is still being followed by some of the worthy citizens of Saskatoon.

ELEVEN

January of 1935 was wickedly cold. For ten days the temperature fluctuated between 40° and 50° below zero, dropping to −52° one memorable morning — especially memorable because, when the thermometer fell that low, schools did not open and we had a holiday.

The dry cold of this second winter was even harder on my mother's sinuses. She made no show of her troubles but, as Rachel remembered, "Sometimes your dad would ask me to stay with her in the evening until he returned from lectures at the university so she 'won't try to push a garden fork up her nose to ease the pain.' "

When she was not wrestling with the demons of her flesh, Helen volunteered her time to one of the several organizations trying to assist the destitute, of which there were many. Some neighbourhoods in Saskatoon could not then have been much better off than the poverty-stricken slums of nineteenth-century England. Although nothing was said about it in the press, every local doctor could attest to the fact that people were dying as a result of malnutrition, if not of outright starvation. Others actually froze to death. Although all children stayed home from school when the temperature dropped to −50°, not a few were forced to stay home most of the winter because they did not have clothing warm enough to withstand more than a hard frost.

Helen helped unpack and distribute bales of used clothing sent from relatively well-to-do Ontario. She also worked as a nursing assistant tending the children of the poor in the overcrowded municipal hospital. Not much about these activities was ever mentioned in my hearing but later, when I asked Helen about those days, she told me, "It was the most dreadful time. It seemed *unbelievable* that people could be so poor and suffer so in a world where others had everything they wanted and to spare. We Thomsons had been staunch Conservatives since the Flood, but that winter your father and I joined the CCF* and I believe I might have become a communist if I had only known how to go about it."

This summation from my gentle-natured mother reveals more about the brutality inflicted upon mostly working people during the Depression than a score of sociological reports could do.

---

*Co-operative Commonwealth Federation — the socialist party born in the west during and of the Depression and the forerunner of the New Democratic Party.

Because almost everything going on around a child seems "normal" to him or her, I did not fully realize the extent of the suffering at the time. Nevertheless, the meaning of the phrase "the Dirty Thirties" was driven home to me on an occasion when I went with Rachel to visit a family from her own district.

These people, a Polish-born couple and two sons of about my age, had, like many others, lost their farm to bank foreclosure and had been forced to seek refuge in the city. They were existing in a gaunt shell of a frame house beyond the end of the streetcar line, and were penniless. We took them a small sack of flour, some sugar, four or five cans of beans, and a slab of bacon, all of which Rachel had bought with her own slender earnings.

I can vividly remember the repulsive smell of the one occupied room, all the space the family could hope to heat with the green poplar which was what they had to burn. It was a stench I would not encounter again until, in Newfoundland in the 1950s, I became interested in the island's staple product, salt cod. I was to learn then that there are many grades of salt cod. The best make good eating — provided the cook knows how to treat the otherwise inedible grey slabs. The worst was called "maggoty fish." It was not only maggoty but frequently rotten. No Newfoundlander in his right mind would eat it; nevertheless, in the 1930s, tons of maggoty cod were bought by the federal government and shipped to Saskatchewan to be distributed to the needy. Most prairie dwellers did not have the slightest idea how to prepare it but, whatever they tried, only dire necessity could have brought them to eat the nauseous result.

Incidents like the visit to the Polish family insensibly increased my awareness of the fearful inequalities which exist between the haves and the have-nots in the human world. I often felt uncomfortable, if not a little guilty, about the well-nourished life I led, as compared to the stark existence being endured by many of my contemporaries. Something was desperately wrong in human affairs, but I did not know what the trouble was, or what might be done about it. Consequently I tried with some success to banish that awful awareness out of my conscious mind. It would return to haunt me throughout my adult life.

We better-off youngsters were not greatly inhibited by winter weather. Wrapped around with woollen clothing, wearing a fur-lined, aviator-style helmet, and equipped with a pair of Indian-made snowshoes, I followed my natural bent. On holidays and after school, I prowled the river bank and the prairie, always accompanied by Mutt and usually by some other member of the club.

On really stormy days, I occupied myself writing pieces for *Nature Lore*, or with my bird studies which were engrossing more and more of my attention. I was attempting to learn not only the identifying marks of every species of Canadian bird but their Latin names as well. I had filled our back yard with feeding stations (bountifully funded by the Beaver Club) and when these were blanketed by an invasion of showy, yellow and black evening grosbeaks, I attempted my first bird photographs using my mother's

Brownie box camera — which reduced their images to minute black dots upon a field of white.

Although birds increasingly preoccupied me, I found time for a new love. I had recently read a book about Alfred Nobel, the inventor of dynamite. The success he had enjoyed from the simple discovery that, by combining notoriously unstable nitroglycerine with finely powdered clay, he could produce a stable and versatile explosive, made a great impression on me. The urge to emulate him grew strong. Perhaps I dreamed of some day endowing the Mowat Prize for Literature, or Peace. Perhaps I simply wanted to make big bangs. Whatever, I was sufficiently enthused to take a large part of my savings and buy a No. 7 Chemistry Set designed for the edification of "young scientists."

It was a severe disappointment. Although it contained many little bottles and round wooden containers filled wih chemicals, these all proved to be of the innocuous variety. No matter what combinations I tried, nothing much ever happened. I could test people's spit for acidity (using litmus paper), or dye one of Rachel's handkerchiefs red with logwood chips, but the set was not capable of producing even a stink bomb let alone an explosion. Nevertheless, it did provide me with a starting point from which I moved on to bigger things.

In late February Rachel sadly told us she would have to return to the family farm and take over the woman's work there because her mother was very ill. The news upset me almost as much as if I had been faced with losing the sister or the brother I had never had. It was very distressing to my mother too. Her sinus troubles were still incapacitating her a great deal of the time so, quite apart from her fondness for Rachel, Helen had come to rely heavily upon her help.

Following Rachel's departure my mother went into a decline. Her doctor diagnosed this as "an attack of nerves" and concluded she should go to Montreal where specialists might at least be able to alleviate her physical distress. In any event, she would escape from what remained of the prairie winter.

She was scheduled to leave by train at midnight on March 9. That evening a farewell party was held at our house. It was well-attended and, as was usual at Saskatoon parties, there was lots of liquor. Soon after dinner I absented myself to the back cellar where I had my chemistry "lab." While revelry resounded through the floorboards overhead, I immersed myself in my most important and exciting experiment to date.

My chemistry set and its accompanying handbook having proved inadequate, I had turned to other sources of information and materials. A few hours well spent in the library's reference section produced much information on the varieties and means of manufacture of high explosives. I discovered that many were quite easy to make and that the requisite ingredients were readily available. Ammonium nitrate, for example, could be had in quantity from any farm supply store, where it was sold as fertilizer. Potassium nitrate, powdered charcoal, and sulphur — the basic ingredients of gunpowder — could be bought cheaply at Pinder's

Drug Store. So could glycerine. Nitric acid was a little more difficult to obtain, but a neighbour who taught physics at high school brought me some from the school chemistry lab. In due course I intended to transform these latter two substances into nitroglycerine.

On the day of my mother's departure, I was testing a formula of my own which I believed might produce something entirely new in the way of explosives. It did not seem impossible to me that Mr. Nobel might soon have to look to his laurels. What I did was mix potassium chlorate, potassium nitrate, and ammonium nitrate, then add a soupçon of fulminate of mercury. This last ingredient was obtained by shaking out the contents of a commercial detonator which had been supplied to me by the son of a railroad construction engineer.

The party upstairs was in full swing when the cellar windows blew out, the door at the head of the cellar stairs crashed open, dense and acrid smoke poured into the upstairs rooms, and there was an almighty BANG!

"It sounded," Angus told me some years later, "as if somebody had dropped a trench mortar shell down the chimney. There was confusion amongst the guests and a lot of good liquor got spilled in the rush to get outside. Naturally I thought of you. I plunged down the cellar steps and met you headlong coming up. You looked as if you were made up to be a nigger minstrel. Face black as the ace of spades and eyes as big and round as if you had just seen a ghost. Maybe you had. Only it was more likely the grim reaper you glimpsed in passing."

Having absorbed several martinis, my mother did not fully comprehend what all the fuss was about. She was rushed next door where she was given a sedative. What with one thing and another, she must have been fairly fuzzy by the time she was tenderly loaded into her sleeping compartment on the train. I wasn't there but, according to my father, her parting words were directed to me.

"Tell the sweet little lamb," she purportedly said, "he mustn't play about with matches."

Next morning my containers of chemicals, together with those test-tubes and bits of glass tubing which had escaped the blast, were reduced to splinters and powder by my father before being deposited in the garbage can.

Amazingly, the explosion did me no serious harm. I lost my eyelashes, eyebrows, and some of the hair from my head and, for a while, suffered from an irritation of the eyes. The damage to the house was also minor: several broken windows and a baseball-sized hole in the galvanized iron hot-water tank. Nevertheless, I now put chemistry behind me.

I took my farewell of Nobel's science a few days after Helen's departure. Very early on a Sunday morning I walked to the middle of the 25th Street Bridge and, just as the sun was rising, dropped an object over the balustrade. Then I put my fingers in my ears. There was only a disappointing "wump" as my *pièce de résistance* — a one-pound tobacco can filled with home-made gunpowder and fused with a cannon firecracker — blew a hole in a snowdrift on the river ice. The wild pigeons who lived under the bridge were only momentarily perturbed, and

Saskatoon never knew it had been bombed by a passing spaceship.

———————————

Enthusiasm for the Beaver Club waned sharply with the advent of warm weather. Most of the boys were infected with the spring craze for baseball and the girls with the spring craze for boys. Abandoned by most of its members, the club and *Nature Lore* quietly expired.

I was immune to the baseball virus and, since the testosterone coursing through my veins had not yet reached full flood, was more or less immune to girls as well. More or less. I sometimes had fantasies about Muriel Pinder, the dark-haired daughter of a druggist who had done exceedingly well for himself during the Prohibition years. Muriel lived in a rococo, strawberry-pink, stucco mansion a block away and was regarded by the boys of the neighbourhood as hot stuff. Alas, she did not regard me in the same way. I wrote her a poem once.

No bird that flies in summer skies,
No mouse that lurks in sacred church,
No fish that swims in river dim,
No snake that crawls on sunny walls
Can stir my heart the way you do
With raven hair and eyes so blue.

She returned this offering by next day's post, with her critical evaluation written across it in purple ink.

"Ugh!"

Muriel may not have had a taste for me but her literary taste was impeccable.

And, I must now admit, that poem was not even truthful. Birds, mice, fish, snakes, indeed any creatures that lived in the wild were of more interest to me than any girl.

That winter I had begun a correspondence with Frank Farley, my great uncle on my mother's side. In 1882, at the age of twenty, Frank had left his family farm in Ontario and gone homesteading in the Golden West, as it was then being called. He broke several hundred acres of virgin prairie near Camrose, Alberta, and farmed them to such effect that, when he retired fifty years later, he was wealthy enough to indulge his lifelong interest in birds.

Frank was a self-made naturalist in the tradition which sanctioned and encouraged the collecting of natural history "specimens" with such avidity that rarities were literally pursued to extinction. Frank's particular passion was for birds' eggs. Over the years he had amassed an enormous egg collection and, by the 1930s, was accounted one of Canada's outstanding ornithologists. He now became my mentor, if at some hundreds of miles removed.

Frank was a child of the Victorian age but I was from a different time. True, I *had* collected birds' eggs and butterflies but spasmodically and only in a small way. The possession of large numbers of inanimate objects that had once been alive brought me no lasting delight, whereas I took much pleasure in collecting information about the way animals behaved in life. Frank may have realized that I belonged to a new school. He suggested that I begin keeping records of spring and autumn bird migrations, that I study and keep notes on nesting birds, and that I compile a list of all the species which lived in or passed through Saskatoon and its environs. Finally, he suggested I might become a bird bander.

Bird banders were volunteer enthusiasts from all over North America. Under the supervision of the U.S. Biological Survey and the National Parks Branch of the Canadian Department of the Interior, they were supplied with sequentially numbered little aluminum anklets called bands or rings, each of which bore the legend *Notify Bio Surv Wash DC* incised in tiny print on its inner surface.

The volunteers caught birds by any and every method they could devise, then, with the aid of long-nosed pliers, clamped a band of the appropriate size on one leg of each captive, which was thereupon released. A record of the date, place, and band number was forwarded to either Washington or Ottawa. There it was put on file to await the hoped-for return of the band from some far-distant place whither its feathered host had carried it before (usually) coming to grief. The purpose of the whole elaborate operation was to help illuminate some of the mysteries of bird migration.

Bird banders considered themselves members of an élite fellowship. Each was empowered by a federal permit which was as difficult to obtain as a passport. The prospect of joining this privileged group excited me enormously. On May 14 I made application. "As required by law," I enclosed two testimonials from prominent ornithologists (Frank Farley and an obliging friend of his). The minimum age required of a bander was sixteen and I was then only two days past my fourteenth birthday, but my handwriting was so atrocious that the "4" on the application form could easily have been mistaken for a "6."

On June 14 "Billy M. Mowat, Esq. of Saskatoon, Saskatchewan" duly received Permit No. 1545, authorizing him to capture migratory birds for banding purposes.

This may not have been the *most* exciting moment of my existence but few have surpassed it. With Permit No. 1545 in my possession, and no longer counting myself merely an amateur naturalist, I had officially entered into the Realm of Ornithology wherein I expected to spend the rest of my mortal days.

Helen returned from Montreal at the end of April. The doctors had failed to cure her facial trouble but the spring warmth had at least dulled the pain. She and Angus resumed their hectic social life, but I knew little of their comings and goings for spring had swept me out of their world.

The one member of the Beaver Club who had remained loyal to its memory and to me was Murray Robb. A year younger than myself, he was a gentle youth with soft brown eyes and a peaceable disposition. He now became my closest pal. Every Saturday and most Sundays, too, we would bicycle many miles into the country then leave our "wheels" and prowl on foot over the prairies and through the bluffs looking for birds. The field notes which I began keeping that spring tell me that, on my fourteenth birthday, we saw "a thousand or so Mallard Drakes and Ducks on the slough near Sutherland. Also lots of other kinds such as Gadwalls, Pintails and Baldpates filling the smaller sloughs and too many Canada Geese to even count feeding on last year's stubble fields."

It was May. The prairies were again awash with life and I was there to revel in the annual rejuvenation of the great plains.

The need to find a refuge near water returned as summer neared. This time the Mowats made no distant forays. We hauled the caravan to a site on the edge of the river six miles south of Saskatoon. The land, which had been virgin prairie not long since, was owned by the Saskatoon Country Club. Its members had built a golf course on the level ground above the river valley, leaving a broad swath of alluvial flood plain along the edge of the river undisturbed. This thickly wooded shelf had remained much as it must have been when native people were its only human visitors. Portions of the narrow track worn by these not-so-distant travellers and their horses along the river route were still discernible. An overgrown clearing close to the river and to a little clear-water spring had doubtless been used by them as a camping place since ancient times.

It was into this clearing that, with the help of a farm horse, we hauled the caravan. Having cleared away the brush, we set up an umbrella tent nearby. This became my private quarters. Then we built a stone fireplace, a trestle dining table, and a carefully concealed privy consisting of a deep trench spanned by a peeled poplar log that served as a seat. At the river's edge we constructed a flimsy wharf — which was washed away by the fast-rising river after every significant rainfall. Near it we raised a platform of logs upon which a ten-foot, flat-bottomed dinghy Angus had built during the winter could weather the floods.

Also by the shore was a brush-roofed shelter for the use of bathers or simple loiterers. Angus called it a "go-down," and it became my mother's favourite retreat. From it one could look out across the gleaming half-mile width of the Saskatchewan. Streaked with sandbars, the river waters rolled past like an undulating satin ribbon. Ancient cotton-wood and balm of Gilead trees towered over all, their shimmering leaves shading and cooling us as we lazed below them in the blistering heat of mid-summer.

This special place became our home away from home for as long as we continued to live in Saskatoon. What follows is a reconstruction from my notes of a day spent there in late June of 1935.

The darn chipmunks wake me at dawn. Four or five of them have made my tent the centre of their world. It hasn't got a floor, so they pop in and out under the walls whenever they please. Mutt won't sleep in here any more because they walk all over him. One has made a nest in a little old tin trunk I found up at the club barn and hauled down here to keep stuff in. She comes right at me with her tail stuck straight up if I even touch the trunk; then all the rest hear her chatter and come in to cuss me out. Mostly they live outside but first thing every morning they gather inside to see what crumbs I've left. And there'd better be some crumbs! If not, they'll go scooting right over my face looking for grub.

Nobody else is up. I pull on my shorts and Mutt joins me from under the caravan and we go down to the river. The sun is just climbing into sight and a little mist is lying along the shore. We wade out into the water and, just for the heck of it, Mutt splashes toward a Great Blue Heron fishing on the nearest sandbank. It goes "Gwaaawk!" and flies off around the next bend where it's more private.

While I'm standing there, knee-deep, one of the bank beavers swims by. He knows me well enough by now so he doesn't even slap his tail. I try to join him but he can swim twice as fast as me and anyway he dives. I stick my head under but the water is so muddy I can't see anything. How does he see where he's going underwater?

When we get back to camp Dad is up and dressed and cooking breakfast on the Coleman stove. He has to go to work in Saskatoon. Mum doesn't, so she sleeps in. He and I eat our porridge at the log table and McPhail,* the big, grey ground squirrel who lives under it, comes and stands on my bare foot until I give him a piece of toast. Dad says I should make him work for his food but what can he do? He is such a lump. Even the Magpies take scraps away from him. They aren't here for breakfast today so I'd better take a look at their nest and see if it's all right. The Hunters Club in town is death on Crows and they live outside

---

*Named by Angus in honour of a pompous member of the public library board.

159

today so I'd better take a look at their nest and see if it's all right. The Hunters Club in town is death on Crows and Magpies, and Hawks and Owls too, and pays kids a bounty for their eggs.

*Dad walks up the trail and I hear Eardlie start to snort and they're off to work. It's time to get my field glasses, my bird bands and my notebook, and start doing my nest check. Today Mutt and me begin at the big cottonwood stub a mile up river. The eggs of the Sparrow Hawk nesting in an old Woodpecker hole there are going to be hatching any day. She's got so tame I can reach in and lift her right off the eggs and out the hole, and she doesn't even scratch. Which she certainly did the first time when I put a band on her leg.** *

I guess wild animals get used to anything that doesn't try to hurt them. At least, some surely do. Now, when I climb her tree in Dead Cow Bluff, the Long-eared Owl sits tight on her nest and just turns her head and looks me right in the eyes as if to say, "You again? What do you want this time?" Or maybe she's looking to see if I've brought along another Star-nosed Mole like the one the caddies killed up at the club last week and I put in her nest.

I check the ten nests I know about along this part of the river bank: 4 Wrens, an Orange Crowned Warbler, a Yellow Warbler, 2 Catbirds, the Sparrow Hawk and a Red-tailed

---

* Very few bands are ever recovered because the birds carrying them do not die conveniently for people to find them. However, on February 5, 1936, a man named Ernest Mica shot a sparrow hawk near Flatonia, Texas. She had a band on her leg and Mica returned it to Washington. It was the band I had placed on the little hawk who had nested in the old cottonwood stub.

Hawk. Mostly their eggs have already hatched. The three young Red-tails are still covered with fuzz but are old enough to roll over on their backs and stick their claws up at me. In a couple more days I'll put on their bands. Today I just count the bits and pieces of about a dozen gophers in and around the edges of their nest. I throw a couple of bits down to Mutt just for a joke but he won't even give them a sniff. He likes his dinner cooked.

We finish with the river-side nests and climb up the bank near the old army rifle range. Scratch Eye Bluff is first and Mutt scurries ahead, leading the way. Last week we met a coyote here and it was just lucky for old Mutt the coyote had other business. Or maybe not. Maybe Mutt could have made friends with him but I wouldn't want to bet on it.

Although this is only a small bluff, about as big as a baseball diamond, it's got two Crows' nests and a Magpie's nest in it. All three families get along somehow. But they always gang up on me and Mutt, though they must know darn well by now we're not trouble. There's also a Loggerhead Shrike's nest in the diamond willows beside the bluff and both the Shrikes like to get into the act too. Scratch Eye can be pretty noisy when they all get going!

It takes us a while to check out all the nests in the patch of nice, fresh, green prairie and bluffs between here and the road. A pair of Kingbirds, Brown Thrashers, 2 more pairs of Crows, 4 of Flickers, one each of Brewer's Blackbirds, Meadowlarks, and a couple of pairs of Vesper Sparrows all have nests in around here. Some aren't doing too well. Something got one of the Sparrows and ate her and her eggs. Something else got into a Flicker hole and smashed all the eggs but one. But Flickers are tough. She'll just go on laying until she's got a full clutch again.

Now we cross the road onto the golf links. Nobody is out playing yet but I see one of the caddies I know, Bruce Billings. He lives on a fox farm between here and town, and we kind of like each other. He gives me a wave and I wave back as I go into Hang-Up Bluff. I call it that because a couple of weeks ago I got hung up there. I had climbed up to a Flicker's hole and put my arm in right to the elbow to count her eggs when the stub I was standing on broke off. It dropped me down with my feet swinging and my arm hooked in the hole. For a while I was scared I'd never get free but then I managed to shinny up far enough to get my arm out. It was pretty sore for a while.

Bat Bluff is next. That's where I really got surprised. I stuck my hand into what I thought was a Flicker's nest and poked my finger into a mess of needles. Or that's what it felt like. But it was a bat, and a big one too. When I hauled my hand out she was still holding onto my finger. It didn't really hurt so much but it made me jump — or I would have if I'd been on the ground. Anyway, she had two baby bats in that hole. I took all three of them back to camp and kept them in a wire mesh box in my tent. The mother bat didn't seem to mind. I opened the cage at night and off she'd go, hunting insects I guess. Every now and again I'd hear a little squeaky noise when she flew back into the tent to feed her young. Then, day before yesterday, she was gone and the babies with her. She must have carried them away one at a time. They were Red Bats which are very rare. She didn't go back to the old Flicker's nest. When I looked at it yesterday a pair of Tree Swallows had taken it over. They've already got their nest half built.

By now it's starting to get hot and Mutt lets me know he wants a drink so we head back for camp. Mum is washing her hair at the Go-down. She wants to know what we've seen and I tell her, and then get myself a glass of milk and some oatmeal cookies I share with Mutt and McPhail. McPhail may be afraid of Magpies but seems to think dogs are just moveable tree stumps. Boy, has he got a lot to learn!

Well, that's the way the day goes. All in all, I look at forty-one nests of eighteen kinds of birds and make notes on what is happening to them all, and band eight young Crows.

Late in the afternoon Mutt and Mum and I go swimming then we row the dinghy out to the sandbars and lay there dozing in the sun. We're so quiet a fox swims over from the other side, lands on our bar and would have walked right between us if he hadn't got our wind. Mutt tries to chase him back to his own side but he has other ideas. Mum and I are still laughing at the sight of the two of them swimming upstream against the current and not going anywhere much, when we hear Eardlie honking. Dad is home and it's time to light the fire and cook supper. He's brought sausages for us to grill and a store-made apple pie.

All in all it's been a pretty good day when you come to think about it.

TWELVE

Helen's trip east reinforced her desire to return permanently to familiar places and familiar faces. She did not nag Angus about this — nagging was not her way — but he was aware of how she felt and secretly shared her yearning for the home of their younger years. A Bay of Quinte Bullfrog born and bred, he was starting to feel as if he were being transformed into a Desert Horned Toad.

Despite (or because of) the dust and drought, his imagination seethed with visions of white-winged ships and rolling oceans, which resulted in the Saskatoon Public Library acquiring one of the outstanding collections of maritime books in all of Canada. He, too, was homesick.

Not being privy to my parents' inner feelings which, I must admit, did not greatly interest me in any case, I was taken by surprise when, in mid-summer, Angus announced it was time for us to pay a visit to the past. I had no suspicion that it was also in his mind to see what opportunities might now exist for him as a librarian in eastern Canada.

The prospect of travelling east did not thrill me. I was reluctant to leave my Eden at the country club, but at least I had the satisfaction of seeing most of the fledglings whose lives I had been overseeing depart from their nests before we also departed.

Leaving the caravan behind (itself now an empty nest), we climbed into Eardlie and on July 27 set out for the east. We reached my grandparents' cottage in the Gatineau Hills early in August and Helen and I settled in while Angus set off to "sniff out the lay of the land."

The cottage and its environs, which I had found so enthralling when I was eleven, no longer excited me. While my mother revelled in long days spent in trivial pursuits (swimming, afternoon teas, and bridge games), I grew increasingly restive, longing to be back in sun-scorched and drought-stricken Saskatchewan. In the space of two years, I had become a prairie boy and, although this second visit to the Gatineaus was hardly an ordeal (there were boats to row, lakes in which to swim, and birds, if only *eastern* ones, to watch), it seemed no more than an interlude in my life.

One pleasant memory was of a ten-day visit by my cousin Helen Fair Thomson from Calgary. My uncle Jack's daughter, she was a devil-may-care little blonde a couple of years my junior who was happy to help me enliven things a little.

Grandmother Georgina Thomson, the doyenne of the establishment, made a ritual of drinking a small glass of sherry every evening before going to bed. "Just a *tiny* sip," she would say apologetically, "to help me sleep, you know." This it certainly seemed to do for her snores reverberated through the flimsy walls of the cottage to such effect that, early on, I had shifted *my* bed to the boathouse. But the "tiny sip" turned out to be mere camouflage.

One day Helen Fair, prowling about where she hadn't ought, discovered that our revered grandmother kept a private stash consisting of two gallon jugs of cheap sherry secreted under her bed.

We told no one. Instead, we helped ourselves to Georgie's sleeping medicine, decanting our tithe into a pint milk bottle, after which we filled the sherry jug up to the proper level with naturally brown swamp water. We kept our bottle hidden in the ice-house and on hot days would sip at the sweet and heady wine while lounging in our bathing suits in the damp, cool sawdust.

Apparently Georgina did not sleep any the less soundly for drinking watered booze. According to my cousin, Grandmother's snoring continued unabated.

Late in August Angus returned to us, disheartened. He had found no suitable opening in library work and, in fact, there were precious few in any other activity, either, since the Depression was continuing to deepen. Nothing of this was discussed in my presence but, as what I viewed as exile from the west drew to its end, my mother became unaccountably downcast and Angus uncharacteristically gloomy.

Mutt and I and Angus left the Gatineau in the later part of August, Helen having elected to remain amongst her own kind a while longer, eventually to return to Saskatoon by train.

We three males drove to Oakville to spend some time with my paternal grandparents and my aunt Jean and her son Larry while Angus reconnoitred the library situation in nearby Toronto.

Since I had last seen them, my grandmother Mary had become an irascible invalid and my grandfather Gill had withdrawn so far into his private world as to be almost unreachable. I sought relief from this gloomy situation by writing a poem in which my antipathy for the urban east, and my love of the west, came through, loud and clear.

> *You come from the city, the fester spot*
> *Where man and nature have drifted apart,*
> *Where limbs decay and bodies rot*
> *In the smoke and the grime and the smell.*
>
> *You've left the noise of man and machine*
> *For the land and the sky where the air blows clean,*
> *Where the loudest sound is the eagle's scream,*
> *And the coyotes' lonely wail.*
>
> *There's times you wish that you'd never left,*
> *You think of your friends and you feel bereft.*
> *So you swing your axe with a hand grown deft*
> *And forget as the city grins.*

When the meadowlark sings you sit enthralled.
You are there when the river ice cracks and growls.
You hear as the wolf chorus barks and howls.
You are one with the voice of the Plains.

I hardly need remark that I had been reading a lot of Robert Service.

Observing how restless I was, Angus one day asked if I would like to go into Toronto with him and visit the Royal Ontario Museum. This was equivalent to being invited to visit a major temple of the gods. I was ecstatic.

Next morning Angus dropped me off in front of the vast museum building, promising to pick me up in three hours' time. With considerable trepidation, I climbed the long flight of entrance stairs and asked a guard where the birds were kept. I followed his directions through echoing halls filled with such fascinating objects as Egyptian mummies and ancient Graecian weapons, until I reached the galleries housing the zoological displays. One of these halls was walled with glass cases of stuffed birds, all of them looking deader than death itself.

The bird gallery seemed totally empty of life until a fair-haired young man came striding briskly along. Seeing me contemplating a case of rigid warblers, he stopped and introduced himself as Jim Baillie, an assistant curator.

"You're interested in birds, eh?" he asked.

I nodded vigorously then told him I was from Saskatoon, whereupon he invited me to come with him behind the scenes into a huge room stinking so powerfully of mothballs it almost took my breath away. It was crammed with floor-to-ceiling metal cabinets in which, Jim told me, more than a hundred thousand "study skins" of Canadian birds were stored. He opened the hermetically sealed doors of a cabinet and slid out one of its many drawers to show me row upon row of stuffed avian cadavers laid out stiffly on their backs.

Every species of Canadian bird was to be found in this room, if not in life, at least in a form that could be closely examined and even handled. I was confronted by such a plethora of riches that I could hardly contain myself. Amused by my enthusiasm, Baillie turned me loose to explore the cabinets on my own.

For the next two and a half hours, I wandered in a dream, unaware of anyone else, including the director of the ROM who, on August 22, 1935, wrote in his journal:

"Today I met a fourteen-year-old boy, Farley Mowat, at the museum. He was searching the museum cabinets for birds he did not know or had not seen in Saskatchewan. Red-headed and freckled, he ought to make a name for himself in natural history. He is a grand-nephew of Frank Farley of Camrose, Alberta."

On September 11 Angus, Mutt, and I finally headed for Saskatchewan. Having delayed well beyond his predetermined return date in the vain hope that something acceptable would turn up in Toronto, my father was now in such a tearing hurry that we arrived in Saskatoon three and a half days later. I think he drove most of each night while Mutt and I slept. All I recall of that mad dash is that Angus bought several dozen chocolate éclairs and the three of us subsisted on these for most of the journey. It was some time before I was again able to look a chocolate éclair in the face.

This sort of thing was characteristic of Angus. He believed that, if one was going to indulge oneself, one should do it in a big way. Soon after we got home he decided to indulge our mutual craving for fried onions. He bought a thirty-pound bag and five pounds of fat bacon, and we had fried bacon and onions for dinner every night for a week. Apart from making us both bilious, this diet induced flatulence of such potency that I was a virtual pariah during my first few days as a high-school student at Nutana Collegiate Institute. I could only thank my lucky stars that I had had the foresight to change my name to William. As it was, I had to bear with "Stinky" for some months.

Because some of the teachers at Nutana were interested in the same things I was, I now began to take an interest in school. Frank Wilson, called Monkey Wilson because of his gnarled little face, taught biology and taught it well. He was also a neophyte wildlife photographer with a passable knowledge of birds. Jelly Belly (I can't recall his proper name) taught mathematics but clearly was not enamoured of his subject and was tolerant of those of us who loathed it. On the other hand, he loved Indian lore and was an expert on the Plains tribes. He could sometimes be sidetracked from geometry into telling us tales about the Blackfeet and others such. Then there was Miss Edwards, my English teacher, who not only refrained from forcing us to memorize poetry

we couldn't stomach, but allowed us to write essays on top-
ics of our own choosing. She was most encouraging of my
own literary bent — besides which she was young, nubile,
and not averse to flirting a little, even with the likes of me.

This was something of considerable moment for, at long
last, my male juices were beginning to flow. "I think the little
bugger's balls are finally coming down," was how my father
described the change in me to Don Chisholm who, it may be
remembered, was a close family friend and my mother's most
devoted admirer. I overheard the comment and was not
amused. My testicles had been down for a long time; it was
just that they had not received any call to action.

I had (as noted) discovered masturbation as early as
1933 but since it hadn't pleasured me much then I had lost
interest in it. Now my interest was vividly revived, something which did not escape Angus's attention and which
doubtless inspired his comment to Don Chisholm.
However, he sensibly did not forbid it. Nor did he refer to
it in the bleak, guilt-inducing sexual terminology of the
times as "self-abuse." In fact, so far as I can recall, he only
once referred to it at all. One morning over breakfast (still
oatmeal porridge and honey), he looked me squarely in the
eye and without any preamble said, "In the army, we used
to call it pulling the pud. Everybody did it. Nothing wrong
with it either. Nothing wrong with *you* playing with
yourself, Bunje, so long as you don't get caught by a minis-
ter, or any of the other old women whose idea of pleasure is
attending a good hanging. So carry on until you can try
something better."

I was speechless with embarrassment and, after this out-
burst, I think he was too. We said not another word to each
other until we met again for dinner that evening. Then:
"Your mother won't be home until the end of the month,"
he told me, "so I've gone ahead and hired a girl to look
after us before we get knee-deep in dirty dishes and soiled
socks. Her name's Louise."

To this day I don't know whether Angus hired Louise
because he thought I was in need, or because he was in
need, or genuinely as someone to clean the house and wash
and cook. Maybe all three.

Louise was a knock-out. Nineteen years old, she was,
in my father's eyes, the epitome of western femininity: "a
buxom, big-boned prairie woman." She had a luscious,
husky voice, gleaming black hair and eyes, and couldn't
have cooked anything more elaborate than a pancake had
you paid her double wages. She couldn't actually do anything much in a housewifely way except make and consume vast quantities of chocolate fudge and, of course, do
the washing.

Alas, she did not long remain with us. My mother
arrived home early in October and a week later Louise
departed. I never knew the rights of her dismissal but many
years later Helen responded to an inquiry of mine with
these enigmatic words: "It wasn't so much what Louise
couldn't do. It was what she could."

I was sorry to see her go for I had become fond of her
and for a long time thereafter she was central to many of my
more satisfying dreams. Yet I always had the sneaking suspi-

cion that I was somehow in competition with my father for her. If so, I never had a chance.

Well, there would be other maids — though not for some time to come. After Louise's departure, my mother announced with unusual firmness that we were better off without a maid at all and that henceforward she would do the housework even if, she could not resist adding, "it quite wears me out."

I lost out on two accounts. Louise was gone from my life. And I found myself spending a lot of time washing dishes, peeling potatoes, sweeping, and even dusting, while Helen was confined to bed or to the living-room couch because of her "poor face." Since I knew my mother actually was suffering I did not much resent having to do housework, but I *did* resent having to assume another of a maid's duties.

Helen and Angus had planned a large cocktail party for mid-December. I was told I would be expected to distribute the hors-d'oeuvres and drinks. Not so bad, except that *I was to be accoutred in a maid's uniform and expected to pass myself off as a genuine female servitor.*

I was appalled. The thought of dressing up like a girl was enough to chill the blood of any boy of my age. On the other hand, Christmas was approaching and I had some faint hope it might bring me a real camera so I could stop wasting time and film trying to photograph wildlife with my mother's Brownie. In the end, I allowed myself to be persuaded by my parents' argument that my imposture was intended to be a practical joke, and therefore would be fun.

It didn't feel like fun but I nevertheless did my duty so well that nobody recognized me. I also did it in such a way as to ensure that I would never again be forced into such ignominy.

What I did was to be so assiduous with the cocktail shakers that I got all the heavy drinkers squiffed out of their minds, and did almost as much damage to the light and moderate imbibers. The guests, in turn, did some minor damage to Professor Morton's house and rather more serious damage to my parents' status in the neighbourhood which, though by no means staid, drew the line at parties attended by Saskatoon's finest, in uniform, with sirens blaring.

This was not really my fault. My father had recently bought a double-barrelled, twelve-gauge shotgun made by the prestigious English firm of Fox. This had been an extremely expensive purchase ("Darn him, he's spent a month's salary on it," my mother noted sadly in her diary) and he was inordinately proud of it. As the party gained momentum, nothing would do but that he and some of his sportsmen friends step out onto the front lawn and test his Fox. The targets were some presumably obsolete plates which the Mortons had left in a dusty carton in the cellar.* Several of these were sent spinning into the air over the river bank while Angus and one or two others took turns blazing away at them.

They were antique but not obsolete, as my father discovered to his cost when the Mortons resumed possession of their home.

Soon thereafter the police arrived. However, Angus had already established himself in Saskatoon as a character of some stature so, after enjoining him to keep the peace, the policemen departed. I think they went with some reluctance for the party was continuing with unabated zeal. Although I went to bed not long after, my duty done, I was sleepily aware of "noises off" until dawn next day.

No. I did not get a camera, but then Angus had not the wherewithal to buy me one, his purchase of the shotgun having temporarily depleted his finances. On the other hand, I never again had to imitate a maid, and a month or two later Helen relented and engaged a real one.

Angus had been a pot hunter since his early youth, shooting ducks primarily for food, as Bay of Quinte men had been doing since European settlement began. He was not a killer by inclination and never relished taking the life of any living thing until we moved to Saskatoon. There he became infected by the pathological dysfunction which is called "sport" hunting. Those afflicted by this disease derive pleasure from slaughtering "game" animals ranging from moose to squirrels, from doves to geese, together with all kinds of "vermin" from wolves to crows.*

Many of Saskatoon's so-called sportsmen seemed still to be in the grip of the killing frenzy which, within human

---

Any creature which is even suspected of preying on game animals or of competing with them for food or living space is considered vermin by sport killers, and treated accordingly.

memory, had led to the extermination of the buffalo and the prairie wolf, together with the virtual extinction or massive decimation of dozens of other species of prairie animals.

That Angus, a man capable of deep compassion, could ever have brought himself to join in this butchery of the innocents for pleasure's sake remains a puzzle to me, despite something he told me in later years:

"It was the *hunt*, you understand. Getting up shivering in the dark for bacon and eggs and a mug of tea, and then the sounds and smells of an autumn dawn. Sheer ecstasy! Though there was this terrible paradox about it because when you pressed the trigger and death leapt forth, the mood of almost unbearable ecstasy was shattered. Smashed. Turned into bloody slush, just like the birds we killed. The hunting was right. The killing was an abomination because it wasn't done out of need."

Those words were spoken a long time afterwards.

Not only did he become an avid sport killer in Saskatoon, he made it his business to turn me into one too. This wasn't too difficult. Nothing could have been more attractive than the opportunity to be buddy-buddy with my father in a shared enterprise. Besides which, Angus was right: the desire to hunt, if not to kill, comes naturally to most young males.

Our first sally was on a crisp November day only a few weeks after our arrival in Saskatoon. We were in search of upland birds, and we were both tyros in a new land. Angus was equipped with a borrowed shotgun and I with a .22 rifle.

We were up long before first light (I never really went to sleep that night) and having piled all our paraphernalia into Eardlie's rumble seat, we drove through the grave silence of the sleeping city into the open plains beyond. In the making of the day as we drove along the straight-ruled country roads, the dust boiled and heaved in Eardlie's wake, glowing rosily in the diffused reflection of the tail-light. Occasional jack rabbits startled us by making gargantuan leaps into the cones of the headlights, or raced along beside us like ghostly outriders.

The fields on either side had long since been reaped and the grain threshed. First frosts had turned the stubble pallid as an old man's beard. Tenuous, almost invisible lines of barbed-wire fences drew to a horizon unbroken except for the shadowy outlines of grain elevators in unseen villages at the world's edge. Occasionally we passed a poplar bluff, already naked save for a few doomed clusters of yellow leaves. Rarely, there was a farmhouse, slab-sided and worn by dust and winter gales.

I suppose it was a bleak landscape, yet it evoked in me a feeling of untrammelled freedom that may be incomprehensible to those who live out their lives in the well-tamed confines of the east. In a state of exaltation, we watched the sun leap from the horizon while a haze of high-flying clouds flared overhead in a splendid flow of flame — the very signature of a prairie dawn.

We drove on with the sun in our eyes, and little Eardlie scattered the dust under his prancing wheels. It was morning and my impatience could no longer be contained.

"When do we find the birds?" I asked.

"Depends on what birds you're after," Angus explained authoritatively. "Today we're looking for Huns" — he used the colloquial name for Hungarian partridge with assured familiarity — "and Huns like to come out on the roads at dawn to gravel-up."

I mulled this over. "But there isn't any gravel on these roads — only dust," I said dubiously.

"Well, so there isn't any gravel," my father replied shortly. "Gravel-up is just an expression sportsmen use. In this case, it must mean taking a dust bath. Now keep your eyes skinned and don't talk so much."

There was no time to pursue the matter. He trod hard on the brakes and Eardlie squealed to a halt.

"*There they are!*" Angus whispered fiercely. "Stay in the car! I'll sneak along the ditch and flush them up."

Although the light was brilliant now, I had seen no more than a blurry glimpse of some greyish forms scurrying across the road forty or fifty yards ahead of us. Angus disappeared into the ditch and for a while nothing moved except a solitary gopher, who stood on his hind legs and stared beadily at me while whistling derisively.

Angus was having difficulties. There had been a bumper crop of Russian thistles that year and the ditch was choked with their thorny, wind-blown skeletons. It was my father's first experience with these demonic plants. "Rather like crawling through the Jerry barbed wire in front of Ypres," he told me afterwards.

But he persevered and suddenly was on his feet levelling

his gun at a whirring cluster of rocketing birds. In his excitement, he fired both barrels at once — and immediately disappeared into the ditch again, the double recoil of a twelve-gauge shotgun being quite as formidable as a right to the jaw.

Unscathed, the flock of Huns flew straight down the road towards me. As they passed overhead, I recognized them for as pretty a bevy of meadowlarks as ever I saw.

Angus came back to the car after a while and we drove on. He steered with one hand while picking thistles out of his face with the other. Not much was said between us.

Nevertheless, our first day afield was not without some success. Towards evening we saw a covey of Huns in a stubble field and Angus managed to kill two of them.

We were a proud pair when we drove home. As we were unloading the car in front of our house, Angus observed the approach of one of our neighbours and held up our brace of birds to be admired. The neighbour, a sportkiller of experience, seemed impressed. At any rate he came running towards us — but only to snatch the birds out of my father's hand and splutter, "For God's sake, hide them damn things! The prairie chicken season don't open for another week!"

———————————

Despite this inauspicious beginning, Angus rapidly learned the ropes. He bought a second-hand shotgun and went hunting almost every Saturday with newly acquired sportkiller friends. He learned the trick of stubble hunting — shooting ducks as they fed on grain left behind by the threshers, and where to find upland birds, as well as how to tell the difference between prairie chicken, Hungarian partridge, and meadowlarks.

When the bird season ended for the year, he remained hotly engaged with his new passion. That winter he spent an inordinate amount of time in his cellar workshop building a set of two dozen mallard decoys. These were works of art, made of laminated pine planks with their centres hollowed out for better flotation, meticulously shaped and painted to imitate the famous Greenhead mallard and his mate. Yet not even this could quiet his compulsive ardour for the hunt and so he took his shotgun apart and carved a new and elegant stock and forend out of walnut imported from Ontario. Thereafter, at home or at the office, he could often be found lovingly polishing the gleaming wood with a mixture of linseed oil and vinegar. Not yet content, he had the library buy several books dealing with the ancient Arabic skill of "blueing" steel and, after much arcane experimenting, re-blued the barrels of his gun. This esoteric exercise earned him considerable kudos from others of the hunting fraternity.

Helen observed all this dedication to Nimrod with tolerance but some petulance.

"It's a fixation," she said. "He could have built the dining-room table and chairs we need in half the time it took to make his silly ducks."

I continued to accompany him on some of his hunting forays but my enthusiasm was waning. I had begun to take as much or more pleasure in watching the ducks and upland birds in life as in shooting at them. A poem I wrote at this time indicates that the killing was making me uncomfortable.

Sport

A flash of flame that flickers there,
A rain of lead that hisses by,
A deafening crash that rends the air,
A wreath of smoke floats in the sky.

A bark from the dog as it gallops past,
A laugh from the man who holds the gun,
The flutter of birds that seek to fly,
The words: "Good work!" when the deed is done.

A ring of feathers scattered round
A quivering pulp of flesh and bone.
A pool of blood on the autumn ground.
A life has passed to the great unknown.

Although I did not show Angus this poem, I think he had begun to realize that the admiring son of the huntsman father was becoming disillusioned with the game. In an attempt to rekindle my enthusiasm, he bought me a twenty-gauge shotgun of my own. I was grateful but would have been more so had he instead chosen to buy me a decent pair of binoculars.

Near the end of November, Angus made a major effort to bring me back into the fold.

The travelling library he had organized had made him acquainted with a number of people scattered about the province. One of these was a Ukrainian immigrant named Paul Sawchuk. Paul owned three-quarters of a section on the shores of an immense slough known as Middle Lake, well to the east of Saskatoon.* One Thursday toward the end of the duck and goose season, Paul phoned my father to advise him that huge flocks of Canada geese were massing on the lake at night and feeding in his stubble fields at dawn.

Angus and I had never hunted Canada geese, which are the ultimate target and supreme trophy of the water-fowler. I am sure he concluded that if we went goose hunting together we could recapture the mutual excitement and camaraderie of our first hunting trips. So he arranged for me to take Friday off from school and we set out to try our luck.

It was a cold journey. Snow already lay upon the ground and the north wind was bitter. We arrived at Middle Lake in the early evening and found a frozen waste-land. Not a tree pierced a bleak void heavy with the threat of approaching snow. The roads had become frozen gumbo tracks that seemed to meander without hope across a lunar landscape. The search for Paul's farm proved long and frigid.

His house, when we found it, was a clay-plastered, whitewashed, log shanty perched like a wart on the face of a frozen plain. It had only three rooms, each with one tiny window, yet it held Paul, his wife, his wife's parents, Paul's

---

A section is one square mile.

seven children, and two cousins who had been recruited to help him with the pigs, which were his main stock in trade.

Paul greeted us as if we were lords of the realm and took us into the bosom of his family. Mutt, who by this time had become a bird dog of some considerable pretensions, refused to be taken. Having sniffed the piggy air about the cabin with ill-concealed disgust, he refused even to leave the car. He sat on the seat, his nose dripping, saying "Faugh!" at intervals. It was not until utter darkness had brought with it the brittle breath of winter and the wailing of coyotes close at hand that he came scratching at the cabin door.

We slept on the floor, as did most of Paul's ménage since there appeared to be only one proper bed. The floor offered some advantages because the air at the lower levels contained more oxygen. At that there was none too much and, since the windows could not be opened, the trickle of fresh air which found its way under the door was soon lost in a swirl of other nameless gases. The wood stove remained volcanic throughout the night, and our lungs worked overtime and we sweated profusely.

At 4:00 a.m. Mrs. Sawchuk cooked our breakfast, which seemed to consist of barley gruel with unnameable bits of pig floating fatly in it. Shortly thereafter, storm lantern in hand, Paul guided us down to the soggy shores of the unseen lake and out onto a low mud spit.

He had earlier dug two foxholes for us at the tip of the spit but now there was ice-encrusted water in the holes. There was also a savage wind out of the north-east and, although it was still too dark to see, we could feel the sharp flick of snow driving into our faces. Paul departed and we three settled down in our holes to await the dawn.

I cannot recall ever having felt so cold. We had found a sack for Mutt to lie on but it did him little good. He began to shiver extravagantly and finally his teeth began to chatter. Angus and I were surprised by this. Neither of us had previously heard a dog's teeth chatter but before long all three of us were chattering in unison.

The dawn, when it came at last, was grey and sombre. The sky lightened so imperceptibly that we could hardly detect the coming of the morning. We strained our eyes into swirling snow squalls. Then, abruptly, we heard the sound of wings — of great wings beating. Cold was forgotten. We crouched lower and flexed numb fingers in our shooting gloves.

My father saw them first. He nudged me sharply and I half-turned my head to behold a spectacle of incomparable grandeur. Out of the storm scud, like ghostly ships, a hundred whistling swans bore down upon us on stately wings. They passed directly overhead not half a gunshot from us. I was transported beyond time and space by this vision of unparalleled majesty and mystery. For one fleeting instant I felt that somehow they and I were one. Then they were gone and snow eddies obscured my straining vision.

After that it would not have mattered to me if we had seen no other living thing that day, but the swans were only the forerunners of multitudes. The windy silence was soon pierced by the sonorous cries of seemingly endless flocks of geese that drifted, wraith-like, overhead. They were flying low and we could see them clearly. Snow geese, startlingly white of breast but with jet-black wing tips, beat past while flocks of piebald wavies seemed to keep station on their flanks. An immense V of Canadas came close behind.

As the rush of air through their great pinions sounded in our ears, we jumped up and, in what was more of a conditioned reflex than a conscious act, raised our guns. The honkers veered directly over us and we both fired. The sound of the shots seemed puny, lost in the immensity of wind and singing wings.

It had to have been pure mischance that one of the great geese was hit for, as we later admitted to each other, neither of us had aimed. Nevertheless one fell, appearing gigantic in the tenuous light as it spiralled sharply downward. It struck the water a hundred feet from shore and I saw with sick dismay that it had been winged. It swam off into the growing storm, its neck outstretched, calling… calling…calling after the vanished flock.

Driving back to Saskatoon that night I was filled with repugnance for what we had done. And I was experiencing an indefinable sense of loss. I felt, though I could not then have expressed it, as if I had glimpsed another, magical world — a world of Oneness — and had been denied entry into it through my own stupidity.

I never again hunted for sport, nor did my father ever try to lead me back to it. Although he continued to hunt, if in increasingly desultory fashion until we left Saskatoon for good, I believe his heart, too, was no longer in it.

FOURTEEN

During the final months of 1935 and early 1936, I was the most assiduous young naturalist Saskatoon had even known or, I suspect, is ever likely to know. On Saturdays or Sundays (sometimes both) I trudged resolutely off into the countryside to look for birds. I would be gone almost every day during longer holiday periods. There were lapses, as when I was confined to bed by a bad bout of flu or when a raging blizzard shut me up at home. Otherwise I went on my self-appointed rounds with a degree of dedication which surely made me brother to the legendary postman.

Between September 22 and March 22 I made thirty-four bird hikes covering more than three hundred miles, much of that distance across the winter prairies in sub-zero weather. On January 28 I hiked all day at a temperature of −25°. I did the same at −45° on February 8, and −40° on

February 21 and 23. This was cold stuff, but not cold enough to chill a passion which, in my sere and yellow years, seems almost incomprehensible.

Why did I do it? Did I really believe I was gathering priceless scientific lore about Saskatchewan birds? I hardly think so.

Was I trying to prove I was indestructible? If so, to whom? Apart from Murray Robb and Bruce Billings, one or the other of whom joined me on many of my hikes, none of my peers *knew* what I was doing. I took care that they didn't find out either for they would have thought anyone who did what I was doing was out of his mind. *A guy walks ten miles over the prairie at 40° below just to look for a bunch of birds?*

If not madness, what was it that impelled me? Could it have been a subconscious yet compulsive urge to break through into the world of the Others, even under the most adverse of conditions?

I cannot tell. But I can describe what it was like.

Saturday, December 22nd
It's the first day of the Christmas holidays and I can't wait to get going. I'm up about six but can't get Mutt up until I pull his rug right out from under him. He's having troubles with his bladder these days so I hustle him outside right away and we both pee in the snow. It's dark and cold — thermometer says ten below — and a dusting of snow still coming down after the big storm yesterday. Perfect for snowshoeing.

Nobody else is up so I make toast and marmalade and drink about a gallon of milk. Then I get the lunch Mum made for me last night. Then put on my heavy woollen jacket that comes down to my knees. I'm already wearing thick, wool britches and long underwear. Put on my toque with my leather helmet over it and the big mitts Mum got made by some Icelanders out at Meadow Lake.

I don't bother to put on my snowshoes in town, so Mutt and I wade through the new snow to the street-car stop five blocks away. The first trolley along is the plow and it's having trouble pushing its way through. A little later a passenger trolley comes along and none too soon because, as Dad would say, it's cold as the mill-tail of Hell. We climb aboard and scrunch up close to the stove. Especially Mutt, who's still only about half awake. There's nobody on the car but us and a couple of Poles going home after clearing snow on the streets all night.

Dawn is coming by the time we get to the last stop, at the Exhibition Grounds. There's a kind of smoky light because of all the clouds overhead that you can't see but kind of feel pressing down. I strap on my snowshoes, pulling the lamp-wick bindings tight around the moosehide moccasins Dad got me from some trapper at Prince Albert. Mutt stands there waiting, lifting one foot after another to keep them warm, wishing he was back in the trolley.

Then we're off, cutting kitty-corner across the City Golf Club out onto the prairie. My trail from last time is buried so deep I can't see it. We tramp along for about an hour until we get to Brucie's place. From half a mile away I can smell the foxes his dad raises.

*Brucie's dog, Rex, hears us coming and sets up a racket. Mrs.
Billings opens the door of their little old frame house and calls us in.
Her Scotch accent is so thick I can't make out what she's saying but
I know she's got a hot breakfast waiting. Mr. Billings has already
gone to the fox pens and Bruce comes out of his little back bedroom
just as I take off my coat. His long, blond hair is hanging all over
his face with just his long nose sticking out. We sit at the kitchen
table, my back right up to the old wood stove, and stuff ourselves on
fresh bread and Saskatoon-berry jam, slices of home-cured ham, and
mugs of tea.*

"Where we goin' today?" Bruce asks.

*"Depends on if it snows hard. If it does, I guess we better
stick to the river bank."*

*"Och, ye're a pair of daft dolts!" his mother says. "One
day ye'll get yersels lost and freeze as hard as they icicles." But
she's smiling. She knows we won't get lost, not with old Rex
and Mutt along to find the way.*

*There's not much snow falling now so we decide the heck
with it, we'll head out north-west away from the shelter of the
river valley into the bluff country where we'll be most likely to
find birds. I start off breaking trail; then, after half a mile, Brucie
takes over. It's hard work. The snow is sure deep and the lazy
dogs just saunter along in the path we make. Sometimes one of
them gets too close and steps on the tail of someone's snowshoe
and trips him up, and then there's Old Harry to pay.*

*Our breath makes white, fleecy puffs and freezes into hoar-
frost around the edges of our helmets. There's a bit of wind out on
the open prairie so we have to stop every little while and take off
our mitts and rub away the white spots on our noses and cheeks.*

At last we get out of the wind in amongst the bluffs behind Henry's place. First thing we flush about a dozen partridge out of a drift so close they nearly hit us with their wings. Rex takes after them but Mutt knows better. He sits with his tongue out, laughing, until Rex flounders back.

We've come to the right place. The storm must have brought a lot of animals into the bluffs looking for shelter. In the next hour we see flocks of redpolls, some pine siskins, some evening grosbeaks, a white weasel, a flock of Bohemian waxwings, and then a prairie falcon.

The falcon comes whistling out of a bluff right on the tail of a grouse that crash-lands into a snowbank and just disappears. Smart grouse! The falcon flies off and I'll bet he's pretty mad. Rex has a run at the grouse as it flaps its way out of the drift.

"Come back, stupid!" Bruce yells.

About a mile more and we come to a bunch of coyote tracks. They're so fresh the dogs take one sniff and start crowding us until we can hardly take a step without bumping into one of them. Brucie is calling them cowards when I see something up ahead on top of a rise. There's some snow falling so I can't see too clearly. We stop and stare until we make out four coyotes standing there staring right back at us.

Brucie lets out a holler. "Git outta there, you mangy bums!" He waves his arms and the coyotes just seem to melt away.

We snowshoe up the rise and right on top, where the wind has kept a patch of stubble clear, is a dead pony. We figure it's a wild one off the ranch at Beaver Creek that got lost and froze to death. It's just skin and bones. It must have been trying to paw out one last bit of stubble when it died.

Bruce says we should cut off the tail because you can sell horsehair to a guy in town who fixes furniture but I say we ought to leave it be.

"Well," he argues, "coyotes'll only tear it up anyhow."

"They got more right to it than us," I tell him. I don't know if he agrees or not but it's getting too cold standing around so we go on.

After a while we know it's time for lunch. Neither of us has a watch but our bellies let us know. We pick a spot out of the wind in a thick little bluff and dig down about three feet through the drifts, using our snowshoes for shovels, until we reach the ground. Then we collect twigs and branches and some stubble straw and light a fire. The space we've cleared is just big enough for the fire and the dogs and us. I cut a couple of green branches with a "Y" at the end of each; push them butt-down into the snow and lay another stick between them and across the fire. Bruce hangs our tea billy from this stick. It's an old five-pound jam can, black with soot. We keep putting snow into it until it's mostly full of water, then we add a handful of tea and wait for it to boil. While that's going on we hang pieces of buttered bannock and bread on sticks close to the fire and put a can of pork and beans with a hole punched in the top so it won't explode, down among the coals. That's lunch, except for frozen cake and biscuits.

A bunch of chickadees come to visit while we eat. They fly right down and one lands on my arm, walks to my hand, and helps himself to a beakful of bannock. The others get so excited it feels like the air is getting thick with chickadees. They sure know a soft touch when they find one.

The fire has burned down and we are slurping the last of the tea with lots of sugar in it when a magpie comes calling. He wags his big, black tail and yells down to us, "Got-any-grub-anygrub-anygrub-anygrub?" Brucie laughs and tosses him a crust. He catches it in mid-air and flops off to gulp it down before some other magpie sees him.

Now the snow starts coming down thicker and it's beginning to blow. Soon it'll be drifting. Time for us to head home. We douse the fire and take off, right into the wind. It's from the south but cold enough to freeze the balls off a brass monkey is what Brucie says.

Pretty soon we can't see more than a few feet and our eyes are gumming up with snow. We reach the road near Henry's place and head along it. The drifts are so deep we can't find the road in places but the dogs know and they take the lead. Near the corner at the crossroads something big and white spins off a telephone pole and dives straight down at Mutt. He yelps and jumps aside. It's a snowy owl — big as a barn it looks! It must have thought for a minute Mutt was something it could eat. When it saw the mistake it back-pedalled up and away and vanished into the drift. I guess times are hard all over, not just for people.

Around 4 o'clock we get to Brucie's place but don't even see it until Rex leads us up the lane. By this time the ground drift's got so thick it wouldn't be smart for me to try and get back to town. Mr. Billings says I should stay the night, and he rings up Dad on the old party line, and it's okay.

We eat stewed rabbits for supper. Then we listen to the battery radio for a while before we go to bed. Me and Brucie sleep under a thick, old feather quilt, and Mutt and Rex sleep crowded together behind the stove. It goes to thirty below in the night and the wind howls like banshees but we're all snug as bugs in a rug.

Although my parents generally approved of my activities, the zeal with which I was now pursuing my ornithological interests sometimes gave them pause.

Not content with trying to find out all I could about the external aspects of birds, I became interested in their internal machinery. Whenever I found a dead one that winter I would bring it home, thaw it out, and dissect it in the seclusion of my room. This could be a messy business, as on the occasion when the bird was an over-ripe prairie chicken. My mother attributed the consequent odour to "unwashed boy," and never knew what lurked for several days in an old pan under my bed.

My parents did, however, know about the woodpecker.

They were giving a dinner party one Sunday in January. It was a small, select party for adults only, very formal. The diners were having dessert when, in the midst of a solemn conversation about King George the Fifth's grave illness, I burst into the dining room, dancing with excitement, and bearing aloft a tin plate.

Since there was apparently nothing to be seen on the plate, Angus thought (or so he later said) I was playing Salome without the head. He had begun to reprove me for interrupting my elders, when I stopped him.

"Dad! Dad! I've *found* them! I've *got* them!"

One of the guests was Bessie Woodward, wife of the owner of Saskatoon's daily newspaper, the *Star Phoenix*. Now she asked politely, "Got what, Farley?"

"The testes of a hairy woodpecker! Just look!"

Whereupon I thrust my offering before her startled eyes. The testes were minute but I produced a magnifying glass so the guests could have a close look. Some people left their desserts unfinished.

Whatever Mrs. Woodward thought about it, her husband must have been intrigued. A week or so later he sent me a note asking if I would be interested in writing a weekly column about birds in the *Star*'s Saturday supplement for young people. This was a four-page tabloid called "Prairie Pals" which enjoyed a tremendous popularity with kids of all ages in those days before comic books.

The demise of *Nature Lore* had left me with no outlet for my writing so I seized upon this opportunity with an avidity which was not entirely untainted. Mr. Woodward had said I might be paid for my work if it proved acceptable.

I went all out. School work was neglected even more than usual. I wrote every day after school, picking with two fingers at my father's portable typewriter. I would have written every evening too but the sound of me clicking away apparently got to Angus and he reclaimed his typewriter after dinner to work on what he hoped might be a novel.

In mid-February I sent off a batch of four pieces then sat back to wait, alternating between gloom and hope. I heard nothing directly from the *Star* but when I opened "Prairie Pals" on the last day of the month, there I was in print, and this time in *real* print. The column was called "Birds of the Season" and an introductory paragraph by the editors informed all and sundry that it "came from the talented pen of young Farley Mowat of Saskatoon."

Now that I was formally launching myself into a career as a newspaper columnist, I had decided to come out of my Farley closet and give Billy the go-by, at least officially. The following week I got a cheque for four dollars — a dollar a column — and a note informing me that I would hence-forth have a Thursday deadline to meet.

Heaven was here — was now! I flung myself on the type-writer, frantic to build up a backlog of columns in case I burned out and died young. Visions of achieving immortality as an author danced in my head. I pleaded with Angus to let me use his machine in the evenings and, delighted with his son's success, he agreed and his novel went into temporary abeyance.

Here are some samples from "Birds of the Season."

No. 1 — Chickadees
(Penthestes atricapillus)

"Chick-adee-dee!" How often have you heard this merry lit-tle call and glanced up to see a tiny black and cream acrobat hang-ing from a branch and gazing at you enquiringly with beady black eyes?... His fluffy little body and cheerful whistle are the essence of Winter and his spring call heralds Spring just as forcefully.

...When an intruder ventures near his nest a string of well-chosen remarks are hurled from an indignant throat and if this fails to have the desired effect of sending you on your way, the midget will attack with true pygmy valour. He flashes his wings in your face with a gesture that cannot be misunderstood and if you still linger he will make dire threats as to your fate.

A trifle overblown, no doubt, but at least innocuous. As I gained confidence, I began to load my pieces with weightier stuff.

The Waxwings

(Bombycilla garrula, B. cedorum)

When the cold north wind rages over the prairies it may whisk before it a cloud-like flock of swift-flying birds, wafted like leaves through the bitter air. Next morning they will be sitting as erect as soldiers on some bountifully laden crab-apple or mountain ash tree, conversing in low, thin whistles as they labour to swallow a goodly number of the wizened berries.

...One day I came upon the nest of a Cedar Waxwing in a caragana bush. I could see the yellow-tipped tail of the sleek-plumaged bird crouched over her eggs, her back to me, and I quietly laid my hand on her. I gently removed her to examine the contents of her nest and she, with an anxious whistle, lit on my offending hand and belligerently ordered me off the premises.

...If this fearless and beautiful bird takes it into her crested head to raise a family in your back-yard, nothing will deter her, provided that cats and thoughtless boys with air rifles and slingshots are kept at a distance. Be assured that you will be amply repaid for any kindly protection against such villains you may give her.

This got me into a peck of trouble with boys in my neighbourhood and at Nutana who used air rifles to shoot birds at every opportunity and carried slingshots in their hip pockets as an indispensable item of wear. In fact, one boy named Donnelly planted a BB shot in my ass as I rode my bicycle down the back lane shortly after this piece was published. "You better keep *your* distance!" was his shouted advice to me.

I was learning the hard way that a columnist's life can be fraught with rue.

Snowy Owl
(Nyctea nyctea)

Gliding on silent wings over the unfathomable stillness of the frozen prairie, a great white bird floats eerily over the desolate bluffs and silent farms lying dark and shadowy below the shimmering Northern Lights. Across the bleak, snow-bound and wind-swept fields, a barely perceptible rabbit bounces with easy effort. As it passes over the unsuspecting hare the great shadow swerves and flits, moth-like, toward the ground. The unbroken silence is pierced by a quickly stifled scream and the shadowy folds of night envelop the last scene of the survival of the fittest.

...The Snowy Owl is one of the largest of the owl family... To the casual observer it appears as a large, earless, white bird, faintly streaked with brown and possessed of the most puzzlingly silent flight around which many fanciful tales are written. What little is known of the nesting of Nyctea seems to prove that it nests only in the Far North, laying its eggs in a hollow in the tundra. When they hatch, the Lemmings, Ptarmigan and other birds and mammals in the vicinity are sorely chivvied, for a young owl consumes enormous quantities of food before it becomes the hush-winged master of the tundra.

...Owls are not at all discriminating about what passes their rending beaks, and they swallow both the hair and the bones of their victims, including their skulls. When the digestive juices have taken all that is digestible the remains are regurgitated in a soggy ball. This is a wonderful provision of nature and might be a

blessing to man if he could learn to do it too.

...The economic status of the Snowy Owl is on the useful side of the line. His food, while in our part of the country, is chiefly of small rodents and occasionally a sick or wounded Hungarian or Prairie Chicken. Hunters should not condemn him for this as he and all the other kinds of owls are only weeding out the unfit of the game birds, thereby leaving a better and healthier race to carry on.

When the Snowy leaves his wild retreat in the Far North do not welcome him here with shotguns and rifles and do not shoot him as vermin but let him live, a kingly bird among birds, fit to occupy the throne of Monarch of the Air.

This piece stirred up trouble of a different kind. Angus came home from the library with a long face after being visited in his office by some of the local hunting fraternity, including prominent businessmen and a member of his own library board. They had made it clear that anybody who chose publicly to defend vermin against the interests of true sportsmen was pretty close to being vermin himself.

Angus told us about it over dinner.

"You know, Bunje, you could be wrong about the hawks and owls being so damned harmless. But you've got a perfect right to speak your mind about it *or* write about it. Don't let anyone ever tell you otherwise. Only...only for the love of God *do* stay away from birth control or anti-temperance league propaganda in future columns."

This was followed by a good-natured word from Mr. Woodward when next I saw him. "Best not write about hunters any more, Billy," he warned me. "They're a touchy lot."

I played it cool in my next few columns. However, when the first twinges of spring began making themselves felt, I got carried away.

I wrote a piece about the Ruddy Duck in which I devoted a long paragraph to an enthusiastic and graphic description of how this agile little bird makes love under water.

Angus had not thought to warn me to leave sex out of my writing. I'm sure he wished he had because when my column reached the editorial office of the *Star* all hell broke loose. Someone — we were never told who — passed it on to someone else in a women's church league, and the fat was in the fire.

Angus and Helen got several letters accusing them of unforgivable laxity in dealing with my religious education and in allowing me to become contaminated by the evils of sex. "This child," wrote one good lady, "will go straight to hell unless he is led back into the paths of clean thought and Godly behaviour. If he is sent to burn forever, it will be your fault!"

Indignation against my piece flared up so rapidly and fiercely that the *Star Phoenix* bowed before it. With some embarrassment, Mr. Woodward showed Angus a letter from a local businessman who threatened to withhold advertising from the paper "if you allow such disgusting prurience to be placed before the eyes of our children." I believe Mr. Woodward was sorry about what he had to do. He at least saw to it that I was paid for the piece that never ran; but no more "Birds of the Season" appeared in "Prairie Pals."

FIFTEEN

During the winter I had become increasingly chummy with Monkey Wilson, my biology teacher at Nutana Collegiate. When he discovered that I knew birds, he began accompanying me on some of my hikes. Monkey wanted to be a nature photographer, and in return for helping him find suitable subjects he taught me the basics of photography. He also promised that when he got a professional camera he would sell me his old one.

Although more than twice my age, Monkey treated me as an equal. Consequently, I was eager to please him, and when he told me he would dearly like to photograph a great horned owl at its nest, I undertook to find him what he wanted.

Ever since Christmas Bruce Billings and I had been making plans for a camping trip during the Easter holidays.

These would fall in mid-April and ought to coincide with the first surge of spring bird migration. During the five days we would spend trekking along the course of the river, I expected to find at least a few birds we had never seen before. I was also hopeful of locating the nest of a great horned owl.

Bruce and I set off from the Billingses' farm on the morning of April 12. We were burdened with packsacks and bed rolls, and the dogs were saddled with backpacks. Mutt accepted his with a kind of stoic disgust but Rex tried to shed his under every barbed-wire fence we encountered.

Spring was late and the frozen drifts had developed an icy crust through which we sank knee-deep, and the dogs belly-deep. To make matters worse it began to snow. We both thought of turning back to the comfort of Bruce's house but neither cared to be the first to suggest a retreat. So on we ploughed until a darkening sky and thickening snowfall gave us an excuse to call a weary halt.

When we got in back of old Henry's place it was getting to be a real blizzard so we decided to build a wickyup* for the night. We couldn't put it on the sheltered side of the bluff because our supper fire might be seen from Henry's and we were scared he'd come and kick us off. The wind got stronger and blew right into the wickyup and covered us with ashes and sparks and smoke. Pretty soon the fire blew out and we guessed we'd better move to a straw stack and burrow into it for the night or we'd freeze to death.

**A low, lean-to windbreak of brush, open at the front and roofed with tumbleweed and straw.*

We'd just got started digging into one when along comes old
Henry waving a storm lantern and telling us to get to hell off his
land.

He is a mean old guy and I'd have packed up and gone but
Brucie told him we'd sic our dogs onto him if he didn't leave us
alone. Fat lot of good that would have done! But the old bugger is
scared of dogs so he cussed us out some and went away.

We chewed cold bannock for supper then the four of us burrowed about half-way into the stack. There was a foot of snow
outside but we were warm and cosy even though the mice kept
sprinkling straw all over our faces, and Mutt thought he heard
wolves sneaking up on us and kept growling all night.

It was pretty cold when we dug out in the morning. You
could see your breath, and the little thaw ponds were frozen over.
We humped our stuff off Henry's place before stopping to build
another wickyup, and cook beans and tea for breakfast. Then we
pushed on toward Beaver Creek and saw a red-tailed hawk and a
few other birds, but nothing much else because it was too darn
cold! Mutt flushed a horned lark off her nest under the edge of a
snowbank. Those birds must be crazy! They start nesting in
March and get buried up in snow after every storm. She had four
eggs and was so tame she almost sat on old Mutt's nose.

At suppertime we camped for the night in a wickyup we built
on the riverbank. I made fried bannocks and Bruce boiled up some
rice and corned beef for us and the dogs in our tea billy. Then we
lay around the fire trying to keep warm and listening for owls.

My books had told me that horned owls were early
nesters. A pair might begin refurbishing an old crow's or

hawk's nest as early as February and by the first week in April three or four creamy white eggs would be ready to hatch. I knew too that the fiercely territorial owls proclaimed ownership of their home bluffs with nightly hootings. It seemed to me that a good way to find a nest would be to take bearings on the direction of these calls and follow up next day.

Somewhere to the south-west of our camp an owl hooted repeatedly throughout the night. We laid sticks on the ground pointing to the source. Next morning we took bearings with my pocket compass.

We walked along the bearing, searching each bluff we came to, and found nothing but some crows nest building. Then, in mid-afternoon, we found what we were looking for.

Brucie spotted it about thirty feet up in a poplar. A whopping big nest that could have been anything, except it had a big white tail sticking out over the edge. I climbed up and, sure enough, it was a horned owl. She fluffed up big as a barrel then flew a couple of yards away and lit in another tree and looked back at me over her shoulder, not very friendly. I peeked in the nest and it had three eggs, one of them cracked by the beak of a baby owl trying to get out. "Old Monkey Wilson's got his nest!" I yelled down to Bruce.

The temperature hardly went above freezing that day and we had about given up on spring. Again we camped in a straw stack, having nearly perished in our wickyup the night before. During the night the weather changed. We woke to brilliant sunshine and a warm wind from the south

— a wind that smelled like thaw and gumbo. Spring had come to our part of the prairies like a blitzkrieg. By noon the temperature was in the high sixties and the world had begun to run with thaw water. By next day the thaw had flooded fields, coulées, and ditches as if an invisible dam had broken somewhere beyond the horizon. We could hardly move about because of water, slush, and mud.

Late in the afternoon the ice in the river went out. The primal thunder as the surging waters burst their winter bonds shook the river bank where we stood watching. A half-mile-wide expanse of ice began to grind, heave, and split into enormous floes, some several feet thick. We knew that, back in Saskatoon, people would be crowding the bridges watching with delicious apprehension as massive ice blocks smashed headlong into the bridge piers.

If a dam appeared to have broken that day, flooding the surface of the world with water, another broke during the night — this time flooding the realm of air with life. A torrent of bird migration had been unleashed and was pouring northward.

Bruce and I lay awake for hours listening to the thrum and whistle of wings overhead. A three-quarter moon was shining and when we aimed my old field glasses at it, we could watch an almost unbroken procession of ducks, geese, and cranes together with uncountable flocks of lesser birds passing across the lunar disc in silhouette. Most were too small and distant to be identifiable but their numbers were astounding. "Looks like all the birds in the world goin' by!" was Bruce's awed comment.

A few days later, after our return home, I wrote this poem.

Across the darkened dome of prairie sky
Toward the home of shaggy northern bear,
Vast flocks of swiftly flying ducks go by,
Beating with weary wings the singing air.

Close followed by a thousand flocks of geese,
Filling the vernal dusk with eerie cries,
They beat their steady way across the east
Into the flickering Light of Polar skies.

Timeless as life, this strong, unflagging flight,
This muted throb of muffled, beating wings,
Awakens echoes in the prairie night
To memories of a thousand bygone springs.

Mr. Wilson was delighted to hear we had found an owl's nest, but when Bruce and I revisited it the following Saturday, it was empty. A few yards from the nest tree lay the wingless body of one of the adult owls. Bruce later learned it had been shot, and its wings and eggs taken by a neighbour's boy anxious to collect the bounty being offered for "birds of prey" by the sportkillers' organizations in the interests of "conserving" game birds.

The bounty money paid for the eggs and carcasses of hawks, owls, crows, and magpies amounted to only a few cents for each, but cash money was so hard to come by that boys, youths, and even grown men scoured the countryside, virtually eliminating the nests and eggs of "vermin" birds in many regions. This was especially true close to Saskatoon, so it now looked as if Mr. Wilson's prospects for photographing a horned owl on her nest were virtually nil.

Then I had an inspiration. About five miles north-west of the city was a large farm owned by a Mr. Redding, an English immigrant and literary man who was something of an eccentric. He and Angus had become friends and we had visited his place which embraced two sections, partly devoted to wheat but mostly in natural prairie pasture on which Mr. Redding raised Texas cattle. Believing in doing things the natural way, he allowed his several long-horned bulls to roam free. In consequence his land was studiously avoided by all who had no business there, including hunters and egg collectors. It occurred to me that his property might harbour an owl's nest.

Keeping a wary eye on the half-wild cattle, Bruce and I searched the farm's bluffs and on April 19 found a horned owl's nest. It contained three newly hatched chicks.

Very pleased with ourselves we went along to Mr. Redding's farmhouse to tell him about it. He was interested in birds, and when we asked if Mr. Wilson could build a blind in the bluff from which to photograph the owls, he readily assented.

On the following Saturday, Angus agreed to drive Mr. Wilson, Bruce, and me to the Redding farm, because Monkey did not own a car and we had to transport a bolt of green cotton cloth, hammers, saws, and a bag of nails, as well as the camera gear.

We found the young owls safe in their nest. Then, to the distress of the parent birds who flitted about the bluff like a pair of gigantic moths, we built what amounted to a tree house in a cluster of poplars about fifteen feet away from the nest tree. Five feet square and twenty feet above the ground, it was constructed of branches around which Mr. Wilson wrapped the green cotton cloth. He cut a hole for his camera lens in the side facing the nest, and the blind was ready.

According to Mr. Wilson, you could hide in the blind and stay there until the owl thought everything was safe. Then, when she came back to her nest, you could take all the pictures you wanted and she would never even know about it.

"He sure must think owls are dumb," Brucie muttered to me when Mr. Wilson wasn't near. "She may not see him but she could see that tent if her eyes were shut; and I don't think she's going to like it."

Now that the blind was finished, Mr. Wilson said he was ready to try it.

"You boys go off for a walk," he told us. "Make a lot of noise when you're leaving. The books say birds can't count — so the owl will think all three of us have gone and she'll never guess I've stayed up here in the blind."

"Okay, Mr. Wilson," I said. "C'mon, Brucie. Let's get going."

We walked about a mile away to a little slough and started looking for red-winged blackbirds' nests. It was a nice day and we forgot about Mr. Wilson until we began to get hungry. Then we went back to the bluff.

Mr. Wilson was sitting on the ground and he didn't look the least bit well. His face was awfully white and his hands were shaking as he tried to put his big, black camera away in its case. The camera looked as if it had fallen out of a tree. It was all scratched and covered with dirt.

"Get some good pictures, sir?" I asked him cheerfully.

"No, I didn't," Mr. Wilson said, and it was sort of a snarl. "But I'll tell you one thing. Any blame fool who says owls can't count is a liar!"

On the way home Mr. Wilson told us what had happened. About an hour after we went walking the owl came back. She lit on her nest and then she turned around and took a good long look at the little tent, which was on a level with her.

Mr. Wilson was busy focusing his camera and getting ready to take the owl's picture, when she asked a question: "Who-WHOOO-Who-WHOOO?" Then she leapt into the air.

The next thing Mr. Wilson knew, the front was ripped right out of the blind and the owl was looking at him from about a foot away.

He accidentally dropped his camera and then, of course, he had to hurry down to see if it was all right. And that was when we got back to the bluff.

This was only a temporary discouragement. Within a few days, the owls had become so accustomed to the presence of the blind that they paid it no further heed. Over the next several weeks, Mr. Wilson was able to take a series of impressive photographs.

Monkey kept his promise. He sold me his old camera for two dollars. It was a primitive little 35-mm machine with only three exposure speeds, and a fixed lens which made it useless for close-ups. Nevertheless, it was infinitely superior to my mother's box Brownie and I loved it dearly.

My birthday that year — my fifteenth — brought with it the most memorable present I ever received.

I had sent copies of "Birds of the Season" to my great uncle Frank Farley, from which he concluded that I was showing promise as an ornithologist. Without letting me know what he had in mind, he made a proposal to my parents.

Every June for the past five years, Frank had made a journey to the subarctic community of Churchill on Hudson Bay. This was a one-time Hudson's Bay Company post which, in 1927, had been selected as the site for an ocean port from which prairie grain could be shipped to Europe. Over the next several years, a railroad was built north across more than five hundred miles of muskeg and spruce forest to service the new port.

Quite incidentally this last great achievement of North American railroading also provided a means for naturalists to reach a unique concentration point on the Arctic flyway of millions of migrating waterfowl and wading birds. Some individuals of many species which flew this route in spring remained on the tundra near Churchill to nest and lay their eggs. The eggs were the magnet which drew my uncle north. He planned to go to Churchill again in June of 1936, and proposed to take me with him as an egg collector.

My parents gave their assent and details were agreed upon by letter. However, Angus and Helen decided to keep me in the dark until my birthday. This was just as well. Had I known earlier what was transpiring, I would have been able to think of nothing else. When at last it was revealed to me, the proposal was as irresistibly entrancing as the prospect of a trip to the moon might be to a youth of today.

Frank was to pick me up on June 5 when his train passed through Saskatoon. Since this would be more than two weeks before school ended, it posed a problem. My parents, bless them, did not mind my missing that much school time but the principal of Nutana Collegiate would have to authorize such a departure from the rules. I do not know if he would have done so on his own. I do know that Monkey Wilson represented my interests to such effect that the day after my birthday he was able to bring word that not only would I be permitted to leave school early, I would also be excused from writing the end-of-term examinations. For that intervention, if for nothing else, I owe him a lifelong debt of gratitude.

There remained the problems of assembling my outfit — and of mastering my impatience until June 6 arrived.

Angus had read widely on Arctic subjects so he was the expert on what I should take with me. The outfit he finally assembled would have better suited a member of one of Peary's polar expeditions but my father had so much fun gathering it all together that none of us had the heart to bring him down to earth. Uncle Frank did that in due course. When I eventually embarked, I left behind such items as a patented Scott-of-the-Antarctic-style tent large enough to house eight men; a sleeping bag as bulky as a small hay rick and guaranteed to keep one warm at sixty below zero; a manual on how to train and handle dog teams; and an ingenious set of interlocking cooking pots which weighed about as much as I did. I suspect that my father may have been secretly planning a polar expedition of his own since, so far as I know, none of the rejects was ever returned to the store from which it came.

As to my impatience: it was somewhat alleviated by the owls we had found for Mr. Wilson — and by one owl in particular.

On May 20 a torrential rain storm accompanied by near-hurricane winds swept over Saskatoon. The following day when Bruce, Murray, and I visited the owls' nest we found it broken apart and on the ground. Near it were the three chicks. Two were dead but the third — the largest — was still alive.

He was about as big as a chicken and his grown-up feathers were beginning to push through his baby down. He even had the two "horn" feathers growing on his head. He looked completely miserable because all his down and feathers were stuck together in clumps and he was shivering like a leaf.

I thought he wouldn't feel like fighting but when I tried to pick him up he hunched forward, spread his wings, and hissed at me. It was a good try but he was too weak to keep it up and he fell right over on his face.

He looked so wet and sad that I got down on my knees and very slowly put my hand on his back. He stopped hissing and lay still. He felt as cold as ice so I took off my shirt and put it over him. Then I carried him out of the bluff so he could sit in the sunshine and dry off.

It was surprising how fast he got better. Murray had brought along some roast-beef sandwiches. He took some of the meat and held it out to the owl. It looked at him a minute with its head on one side, then gave a funny little hop and came close enough to snatch the meat out of Murray's fingers. It gave a couple of gulps, blinked its eyes once, and the meat was gone.

After that we were friends. When we started to walk away from him, just to see what he would do, he followed along behind us like a dog. He couldn't fly, of course, and he couldn't walk any too well either. He kind of had to jump along. I think he knew he was an orphan and if he stayed with us we'd look after him.

When I sat down again he came up beside me and, after taking a sideways look into my face, hopped up on my leg. I was afraid his big claws would go right through my skin but they didn't hurt at all. He was being very careful.

"Guess he's your owl, all right," Bruce said, and I think he was a little jealous.

We carried him home in my haversack. He didn't like it much but we left his head sticking out so he could at least see where he was going. We put him in the summerhouse and, when Dad got home from work, the owl was sitting in there on an orange crate.

"What are you going to call him?" my father asked.

*I hadn't thought of a name up until that moment. Now
I remembered Christopher Robin's owl in Winnie the Pooh.*

"His name is Wol," I said.

And Wol he was, forever after.

SIXTEEN

On June 5, according to Helen's diary, "Bunje up at 3:00 a.m. No peace for any of us until 7:30 when Uncle Frank's train arrived and we went to meet it. Bunje terribly excited."

That I was. When Uncle Frank's rangy great frame swung down from the steps of the parlour car, I could hardly have been more agitated if God himself had alighted.

I had not previously met Frank in the flesh and he certainly seemed bigger than life. He stood a lean six feet three inches tall in knee-length, lace-up boots. His head, under a soft felt hat, was a mountain crag dominated by the famous (in the family) Farley nose. He had the washed-out stare of a turkey vulture. All in all, he was the most intimidating figure of a man I had ever encountered.

Now he was smiling. One ham hand swept down and gripped my shoulder so powerfully I wanted to squeal like a puppy that has been stepped on.

"So this is the bird-boy, eh?" Frank boomed, shaking my seventy-five-pound frame none too gently. "Not much bigger than a bird at that."

He let me go and turned to introduce a slight, dark-haired young man who had descended behind him — Albert Wilks, a twenty-year-old school teacher who had also been signed on as an egg collector.

Tipped off by Angus who was always *au fait* with the press, a reporter from the *Star Phoenix* was on hand to interview the "world-famous ornithologist." The interview was conducted over breakfast in Wang's Chinese Café where, for the first time, I heard what Frank had in mind.

We would camp on the tundra near Churchill, he told us, until the pack ice covering the inland sea called Hudson Bay slackened enough to let us travel in sea-going canoes accompanied by two Barrenground trappers, "Eskimo" Harris and "Windy" Smith, north up the coast to the Seal River. We would then set up a base camp and spend a month on and around the Seal, making the first scientific collection of animal life from the region. It was to be hoped, Frank added portentously, that the collection would not only encompass birds' eggs but would also include white wolves, Arctic foxes, and seals.

My breakfast went untouched. I was so bedazzled by visions of what lay ahead that I may have been slightly catatonic by the time my parents saw the three of us aboard the noon train. My mother thought I looked angelic, but stunned would probably have been closer to the truth.

By dinner time the train had left the "big prairie" behind and was running north and east through poplar and birch parkland. At midnight it came to a halt beside a cluster of shacks and a small station which bore the tantalizing name "Hudson Bay Junction." Here we disembarked with all our gear to await the arrival of a train from Winnipeg which would take us on to The Pas.

At 3:30 a.m. a baleful whistle roused me from broken sleep on a station bench. We stumbled aboard and found ourselves in the nineteenth century. Our chariot to the North was a colonist car built in the 1880s to ferry European immigrants westward from Montreal after they had been disgorged from the bowels of trans-Atlantic sailing ships.

Colonist cars were designed to transport the impoverished at minimum cost. No effort had been spared to preclude anything smacking of comfort. The seats were made of hardwood slats. They faced each other in pairs and could be slid together in such a way that each pair formed a crowded sleeping platform for four people. There was no upholstery of any kind, no mattresses, and no cushions. Lighting was provided by oil lamps whose chimneys were dark with age and soot. Our car was heated by a coal stove upon which passengers could boil water for tea and for washing, and do their cooking. Some colonist cars had flush toilets of a sort but not ours. There was only a hole in the floor of the toilet cubicle through which one could watch the ties flicker past — an exercise that gave me vertigo.

The car was only half full when we boarded so we were able to claim a two-seat "section" for ourselves. We were lucky. At The Pas — the last settlement en route to the Arctic — the car would become so crowded that some people would have to lay their bed rolls in the aisles.

Our fellow passengers were mostly trappers of European, Indian, and mixed blood, accompanied by their women and children. We also had two Roman Catholic missionaries, and the engineer and three crew members of a Hudson's Bay Company schooner which had spent the winter frozen in the ice at Churchill. All these were exotic enough, but most fascinating was a trio of Eskimos on the first lap of a long voyage back to their homes in the high Arctic after having spent many months in a tuberculosis sanitorium in southern Manitoba. They spoke no English and, since nobody else in the car spoke Inuktitut, I could not begin to satisfy my enormous curiosity about them.

We reached The Pas at noon. Despite its curious name, it was no more than a ramshackle little frontier village serving as the southern terminus of the Hudson Bay Railway which, all in its own good time, would carry us to Churchill. The northern train was made up of a long string of wheat-filled boxcars to which our colonial car, a baggage car, and a caboose were appended like the tail of a dog.

At dusk we pulled out of The Pas and began the long haul northward at a sedate twenty miles an hour — a speed we were never to exceed and which we often fell far below.

By now we had entered the true boreal forest and were bumping along through a seemingly endless black spruce shroud, broken here and there by quagmires and little ponds. Frank joined me at one of the dirt-streaked windows as I looked out upon a broad sweep of saturated "moose pasture" thinly dotted with tamarack trees.

"That's muskeg, my boy. You'll see enough of it before we're through. Fact, most of what you'll see from now on until we reach the edge of the Barrens is just like this. That's why the train is called the Muskeg Express. By some. Some call it the Muskeg Crawler and claim you could walk the five hundred miles from The Pas to Churchill quicker."

Our home on wheels now began to come vigorously to life. Someone stoked up the stove with billets of birch and soon the aroma of bannocks frying in pork fat assailed us, mixed with the molasses-laden reek of the "twist" tobacco most trappers smoked. Those who did not smoke either chewed or used "snouse" (Copenhagen snuff). There were no cuspidors and one would not have wanted to walk about barefoot.

Blackened tea billies came to the boil and were passed along from seat to seat so that everyone could fill his or her mug with a smoky brew heavily laced with sugar. Bert heated us up a pan of pork and beans. A Cree woman across the aisle suckled a young baby at her breast while feeding an older one canned milk out of a beer bottle...and I stared until my eyes bulged.

This being the first night out of The Pas, there was a considerable celebration. Lusty songs were sung in Cree, French, English, and unidentifiable tongues. Bottles were freely passed around. Some of the men began playing cards and there was a brief fight during which I thought I saw the flash of a knife. The noise level mounted by the minute.

Two young women began having a shrill argument over possession of a very hairy, very drunk young man.

At this juncture one of the train men (there was no conductor) came along and leaned down to yell something in Frank's ear. My uncle nodded and pulled me to my feet. "Get your bed roll!" he bellowed.

We swayed to the end of our car, passed through the baggage car (which contained several canoes and a line of Indian dogs chained to a cable along one wall), then we were in the caboose.

"You'll sleep here," Frank told me. "It'll keep you out of trouble. And it'll be a damn sight quieter."

So it was, but much duller. Although I had a bunk and mattress to myself, I regretted missing what might be happening in the colonist car.

Next morning the train crew shared their breakfast with me and the brakeman made me free of the cupola. Reached by a short ladder, it offered an unparalleled view of the country we were passing through. It was like having one's own observation car. I could also step out onto the little porch at the rear of the caboose. I was having a pee from this vantage point when something flipped up from the road-bed and went singing past my head. Another missile whirred by and I hurriedly stepped inside. I told one of the trainmen what had happened and he laughed.

"That's the spikes popping out. You see, kid, the road-bed over the muskeg is so spongy the tracks sink down under the weight of the train then, when they spring back up, they flip loose spikes out of the ties just like stones shot out of a slingshot."

I gave the back porch a miss thereafter and used the inside facilities, unattractive though they were.

I spent a lot of time in the cupola looking for wolves, moose, deer — whatever the vast spruce forests and muskegs might have to offer. They had very little. I saw one solitary moose lumbering away from the track, and an occasional raven.

That was about all, except for human beings and they too were scarce. Occasionally the Muskeg Express would ooze to a halt in the middle of nowhere and a couple of human figures would emerge from the forest to take delivery of packages tossed out of the baggage car.

Sometimes when the train stopped there would be nobody and nothing in sight but trees, until one of our passengers shouldered his pack and climbed down from the car to set off for his trapping cabin in the back of beyond.

Sometimes a canoe would be unloaded and a family of Crees would leave us to paddle away in it. For the rest, there were only the section points, spaced about fifty miles apart. At each of these, two or three section men charged with track maintenance lived with their families in red-painted cabins under grandiose station signs that read: Wekusko...Wabowden...La Pérouse...Sipiwesk...

During the morning of the second day out of The Pas, we crossed the mighty Nelson River flowing eastward into Hudson Bay. The right-of-way now pointed due north and the train ran — crawled, rather — on a road-bed that literally floated on muskeg. The muskeg in turn floated on permafrost — the eternally frozen underpinnings of a land which, even in the first week of June, was still cross-hatched by huge snowdrifts and whose lakes and major rivers were still frozen. According to Uncle Frank, spring was very late this year and he grew gloomy about the prospects for travelling on Hudson Bay.

The uncertain footing now slowed the Muskeg Express to something less than a crawl and there was little to see of interest in the snow-striped landscape. I tried entertaining myself by clocking the slow passage of the black and white mile boards nailed to telegraph poles. By Mile 380 I had tired of this game and was reduced to reading a book. It appeared that nothing was going to happen until we finally reached Churchill.

But at Mile 410 something *did* happen. I had earlier noticed that the succession of stunted spruce trees was being pierced by openings running out of the north-west. When I asked Uncle Frank about these, he explained that they were fingers of tundra thrusting southward from the vast Arctic plains which comprise the Barrenlands.

I went back to the cupola with renewed interest and had just seen Mile 410 slide slowly past when the rusty whistle of the old engine began to give tongue with a reckless disregard for steam pressure. At the first blast I looked forward over the humped backs of the grain cars.

A flowing, brown river was surging out of the shrunk-en forest to the eastward, plunging through the drifts to pour across the track ahead of us. But this was no river of water — it was a river of life. I had my field glasses to my eyes in an instant and the stream dissolved into its myriad parts. Each was a long-legged caribou.

"*C'est la Foule!*" The French-Canadian brakeman had climbed up into the cupola beside me. It is the Throng! This was the name given by early French explorers to the most spectacular display of animal life still to be seen on our continent or perhaps anywhere on earth — the mass migration of the Barrenland caribou, the wild reindeer of the Canadian North.

The train whistle continued to blow with increasing exasperation but the oncoming hordes did not deviate from their own right-of-way, which clearly took precedence over ours. They did not hurry their steady, loose-limbed lope. At last the engineer gave up his attempts to intimidate this oblivious multitude. With a resigned whiffle of steam the train came to a halt.

For an hour that river of caribou flowed unhurriedly into the north-west. Then it began to thin and soon was gone. The old engine gathered its strength; passengers who had alighted to stretch their legs climbed back aboard and we, too, continued north.

The dwarf trees began to march along beside us again but I did not see them. I was intoxicated by the vision of the Throng. Many years later it would inexorably draw me back to the domain of the caribou.

We rolled sluggishly into Churchill at 11:00 p.m. — and it was still broad daylight, for we were now in the Land of the Long Day at a latitude not far short of the southern tip of Greenland. With one exception there was not a great deal to catch the eye. Winter still held the place in thrall. A sprawl of unpainted clapboard shacks and shanties lay nearly buried in drifts which were successfully resisting the half-hearted onslaught of a belated spring. The whole vast sweep of Hudson Bay stretching to the northern and eastern horizons was still ice-bound. So was the broad Churchill River, although its tidal estuary displayed a frigid mixture of open water and swirling floes. A treeless waste of tundra composed of frozen mosses, lichens, peat bogs, and little ponds surrounded the bravely named Townsite. The sky was sombre and a dusting of snow hung in the chill air. It was a scene in which modern man did not seem to belong, yet it was dominated by a man-made object.

A gargantuan concrete grain elevator loomed monstrously over the surrounds of Churchill, appearing even more enormous than it actually was in a landscape where every other sign of human life seemed puny. Towering fifteen storeys tall, this monolith, with its adjacent storage silos and associated docks for ocean-going vessels, was the reason the Hudson Bay Railway existed.

Begun just before the Crash, the complex had been intended to make Churchill a shipping port for the hundreds of thousands of tons of prairie grain annually destined to eastern Atlantic and Indian Ocean ports. It was a visionary mega-project which, like so many such, turned out to be an economic disaster. Yet when I first saw it that grey day in June of 1936, I thought it was something to rival the pyramids — one of the man-made Wonders of the World.

Perhaps it was. But before I had been many days in its shadow, it had ceased to engage me. By then I had become enchanted by the wonders of another world — where man's works played no part.

The morning after our arrival we loaded our gear on a hand jigger — a little rail car propelled by manpower. With Uncle Frank and Bert pumping the handles we rattled out of Churchill on a narrow-gauge spur line. Our destination was an abandoned construction shack standing in lonely decrepitude on the bald tundra some eight miles south-east of Churchill. Shanty-roofed, with tar-paper walls, it contained a rusty barrel stove, two double-tiered bunks, a broken table, and not much else except the frozen corpse of a white Arctic fox that had apparently jumped in through a broken window and failed to find its way out again.

We settled in to await the withdrawal of the pack ice from the coast of Hudson Bay. As it happened, the ice never did withdraw while we remained in Churchill, so we stayed on at the Black Shack, as Bert named it, for the duration.

"At" but seldom "in." Uncle Frank would have made (and maybe was in a previous incarnation) an effective slave driver. Since the nights never got wholly dark, he regarded sleeping as a waste of time.

"Look about you," he lectured me as I tried to lie abed one shivering morning when our water pails had an inch of ice on them. "The birds out on the tundra haven't slept a wink. Too much to do! Too busy! And here it is 4:00 a.m. and you want more sleep! Up and at 'em, sonny!"

Bert was our cook. Burned cornmeal porridge was his specialty but we also ate canned beans; bannock spread with molasses; fat bacon and, on special occasions, cornmeal mush that had been allowed to solidify overnight before being sliced and fried in bacon fat. Frank explained that he had intended us to "live off the land" at Seal River: "Lots of seal meat; maybe a haunch or two of caribou." But since there were neither seals nor caribou where we were, Frank spent much of his time roaming the surrounding tundra blasting ducks and ptarmigan with his double-barrelled shotgun. Bert made watery concoctions that he called Mulligan stew from some of these victims of my uncle's gun, but the corpses of many ended up in a nearby ditch which served as our garbage dump.

I agonized a little about this apparently wanton killing of breeding birds at the peak of the nesting season until Uncle Frank put me straight.

"Don't be soft, boy. There's millions more out there. We're doing this for science. I measure every specimen I shoot and note the condition of its plumage. Science needs to know these things."

Years would pass before I would realize that collecting expeditions such as ours were little more than high-grade plundering operations conducted in the hallowed name of Science. However, for the moment my qualms were stilled and I could go about my duties with an easy conscience.

My duties were straightforward enough.

"Find every nest you can," Frank instructed Bert and me. "The rarer the bird the better. If the nest hasn't got a full clutch, mark the spot and leave it 'til it has. If you aren't sure what species it is, shoot the parent bird and bring it back when you bring the eggs."

With half a dozen tobacco cans filled with cotton wool in our haversacks together with our lunches, Bert and I would be out every day and all day, unless it was pouring rain or, as happened sometimes, snowing so hard that searching for nests would have been useless. I was a good nest finder and I loved the work. When I flushed a rarity, such as a Hudsonian Godwit, and found four eggs ready for the taking, I would feel as elated as if I had found four gold nuggets.

So many waterfowl and wading birds clustered on the tundra that there seemed hardly room enough for all of them to nest. As I sloshed across the still-half-frozen morass of water and mossy tussocks, curlews, several species of plovers, many varieties of sandpipers, and numerous kinds of ducks would rise before me, filling the air with their cries of alarm.

I took a heavy toll from their nests.

Having made my way back to the shack for supper, dog-tired and, like as not, soaking wet from falling through the rotting ice of a pond, I would spread my day's "take" on the table to be admired. Once I unpacked thirteen clutches from my tobacco cans, bettering anything Bert or Frank himself had so far collected in a single day. Frank rewarded me with kind words: "You'll make a good scientist, my boy."

According to my uncle, one egg by itself had no scientific value so we always took the full clutch, thereby

ensuring that, because the season was too short to allow
the birds to nest again, the adult pair would raise no young
that year. The real truth of the matter was that the eggs
had no *commercial* value unless a whole clutch could be displayed as a unit in a collector's glass-topped case. This was
something else I was still to learn.

To blow an egg we forced air through a pipette into a
small hole bored in its side, whereupon, if fresh, the contents would come bubbling out. If the egg was heavily
incubated, we would have to delicately draw out the
embryo, piece by bloody piece, using a needle with a bent
tip. We saved the contents of fresh eggs and those only
slightly incubated, for omelettes which we ate as bedtime
snacks. I remember one such made from Arctic loons', Old
Squaw ducks', and a mixed lot of shore birds' eggs; it had a
distinctly pink tinge and a meaty flavour, probably because
the incubation season was by then well-advanced.

Lemmings were all around us, both inside and outside
the shack. It was a peak year in their seven-year cycle of
abundance and they were making the most of it. Friendly
little creatures looking not unlike small hamsters, they
would sometimes crawl across my lap as I sat on a tussock
eating my lunch. They would also run all over the cabin
floor, paying no heed to us until Bert lost patience and tried
to sweep them out the door.

Egg collecting was not all beer and skittles. One morning
the sun shone, the snow was melting, and it really felt like
spring so the three of us set off together to explore the wall of
granite which fringed the still-frozen bay like a titanic dyke.

We were after the eggs of rough-legged hawks (famed lemming hunters) who occupied a chain of nests built at half-mile intervals on ledges along the seaward face of the dyke.

According to my uncle these nests were ancestral possessions used by generations of rough-legs. Not all were occupied every year. The year after the lemming population "crashed," at the bottom of its cycle, the hawks might use only every second or third nest and, instead of laying a clutch of four or five eggs, might lay only one or two. They were able to adjust their reproductive capacity to the available food supply, something human beings seem incapable of doing.

Because 1936 was a good lemming year every nest held a full clutch. Bert and I had to gather these by scaling the face of the wall or by descending from above. Either way it was a risky business. We had each delivered two clutches to Frank waiting below on the ice-cluttered beach when the hawks decided things had gone far enough.

The ones we had already robbed had been following us, shrieking their distress as they soared overhead. Now, as I began to ascend to my third nest, they began to descend. One by one, like a squadron of attacking fighter aircraft, they stooped on me, talons outstretched and beaks gaping wide. The first one missed by no more than a foot and made me cower against the cliff wall. The second struck home.

My head was buffeted hard against the rock by fiercely beating wings. I raised an arm to protect myself and it was raked from wrist to elbow by sharp talons. For a horrible instant I thought I was going to fall; then Frank's shotgun bellowed and my attacker spun away, still screaming defiance.

I did not wait for the next attack but slid down the face of the cliff to land, scared and shaking, on the beach. Bert bound up my arm with his handkerchief but my uncle offered scant sympathy. He was eyeing the circling hawks and the nest which still contained the eggs he coveted.

"You must have done something to upset them," he said crossly, which surely had to be the understatement of the year.

To do him justice, Frank tried to make amends. When next he pumped the jigger into town for supplies and to see if any possibility existed of making the voyage to Seal River, he declared a holiday and took me along.

We went first to "Ma" Riddoch's Hudson Hotel, a dark little cavern of a tavern to which a few rooms for guests, or girls, or both had been attached. Here I met the redoubtable "Eskimo" Charlie and listened while he and Frank had several beers. Charlie was a weather-beaten Scandinavian with beetling brows who was reputed to have had several Eskimo wives. He certainly did have an awesome store of foul language. There was no effing hope of getting to the effing Seal, he told us, but he would be happy to take us hunting effing white whales in the effing estuary — where the whales were gathering to calve. Frank could see no profit, scientific or otherwise, in such a venture and turned it down.

Later that day I went alone to the docks and watched several hundred belugas, as they are known in the North, disporting themselves in the estuary shallows…until a pair of motor boats put out from shore and their crews began to blaze away with heavy-calibre rifles. What followed was a marine version of shooting goldfish in a bowl. The whales churned the shoal water into foam in their efforts to escape but I could see splashes of crimson appearing on the backs and flanks of many of them.

When I told my uncle about it, he was disapproving. "That'll be some of the men employed at the elevator having a little sport. Foolishness! The Chips might use a couple of beluga for dog feed but most of the ones those fellows hit will roll up on the beach after they die and stink the place up come summertime."

The "Chips" were Chipeweyan Indians from the interior of northern Manitoba who made their way out to Churchill by dog team each spring before the thaw. They came to trade pelts to the Hudson's Bay Company post and were of interest to Frank as a possible source of specimens. We visited their camp and, while he dickered for a pair of white fox pelts, I looked about with awe not untinged with apprehension.

These were the most primitive people I had ever seen. Small, dark, solemn (in our presence), they spoke a peculiar language full of rustling sibilants. They were partly dressed in caribou-skin clothing which was shedding and looked singularly ratty. About twenty of them were living in squat little teepees of soot-blackened canvas full of rips and patches. When I came too close to one of these, an old woman shook her gnarled fist at me, but a much younger woman — hardly more than a girl — smiled and beckoned to me while at the same time opening the front of her shirt. I did not know whether she was being seductive or merely mocking my youthfulness. I retreated in confusion and took shelter behind my uncle.

He had bought his fox skins and was now being offered something so exotic I could hardly contain my excitement. A live wolf pup! When he shook his head at the rather pathetic little animal straining away from us on the end of a dog chain, I couldn't contain myself.

"*I'll* buy it!" I cried urgently. "I'll take it home for Mutt! Please, Uncle Frank, tell them I'll buy it!"

He continued to shake his head. "You can't afford it. They want its bounty value and that's five dollars."

"Lend me the money," I pleaded. "I'll pay it back. I promise!"

"You're talking foolishness, boy. Come along. We're going back to camp."

Summer exploded during the last week of June. Overnight the temperature soared into the sixties and, but for the hordes of mosquitoes emerging from the tundra ponds, we could have gone around naked. The remaining ice and snow shrank visibly before our eyes. The egg-laying season was coming to an end and soon it would be time for us to leave.

There were some chores to be attended to first. Foremost of these, as far as I was concerned, was collecting one more set of rough-legged hawk's eggs to compensate for the clutch I had failed to get. Frank had said nothing directly to me about this but I knew it was on his mind.

One warm and sunny morning I set off to the coastal ridge. I did not go back to the stretch we had already robbed but headed farther east. Nests proved few and far between in this area and it was not until afternoon that I found one, in a cleft of rock fifty feet above the beach. This time I climbed down to it from above, keeping a wary eye on the parent birds wheeling disconsolately overhead. They did not attack and I stole their eggs, wrapped them in cotton wool and packed them in my haversack. That done, I looked about from my high vantage point.

Below me the waters sucked and seethed at stranded ice floes along the shore. Open-water leads criss-crossed the decaying pack to seaward. I looked eastward along the beach and saw a large reddish object. Through my field glasses it revealed itself as the shattered remnant of a ship.

Few things will fire a boy's curiosity as hotly as a wreck, and I hurried off to examine this one. It consisted of the forward half of a small coastal freighter which must have driven ashore many years earlier. I climbed through a maze of twisted, rusty plates until I was standing high on the angled rise of the bow.

And then discovered I was not alone!

Not more than a hundred yards away, three ivory-white bears were ambling unconcernedly towards me. The leader of the trio seemed unbelievably huge, though its followers were not much bigger than a pair of spaniels. I did not need to be a naturalist to know that this was a female polar bear and her cubs.

I was terrified. "Always stay clear of a sow bear with cubs" was a maxim of which I was well aware though, in my case, it had referred to the relatively small black bears of more southern climes. I felt that the warning must apply in spades to the monstrous apparition now padding in my direction with such fluid grace.

I thought of fleeing but to move would have meant revealing myself — and I had no stomach for a race with her! The light breeze was in my favour, blowing towards me, so I could hope the bears might pass by without ever realizing I was crouching in abject fear ten feet above them.

They were within a dozen yards when, for no apparent reason, the female abruptly stopped and reared back on her ample haunches while extending her forelegs for balance. Her immense paws hung down before her, revealing their long, curved claws. Perhaps I moved, or maybe she heard my heart pounding. She looked up and our glances locked. Her black nose wrinkled. She sniffed explosively then, with an astonishing lithesomeness for so huge a beast, slewed around and was off at a gallop in the direction from which she had come, her pups bounding along behind her.

My own departure was almost as precipitate. I fled so fast that the eggs in my pack had become the ingredients for another omelette before I reached Black Shack.

———————————

A week later we again boarded the Muskeg Express. The first of a long series of visits I would eventually make to Churchill and on into the mighty sweep of tundra stretching northward from that Arctic gateway was at an end.

The journey home was an anticlimax. No reporters were waiting at the Saskatoon railway station to welcome the returning explorer. There was only Angus, and he seemed preoccupied.

Having greeted Frank and Bert rather perfunctorily, he turned to me.

"Thank heavens you're back! I'm sick and tired of being foster-parent to an owl. Mutt's threatening to leave home and your mother's about to have an attack of nerves. You and your damned pets!" He paused as he caught sight of the slatted crate I was half-hiding behind my back.

"Ohmigawd! What have you brought back *this* time?"

What I had in the crate was a long-tailed jaegar, a large and strikingly patterned Arctic bird which looks like a gull but acts like a hawk. I had rescued this one after Uncle Frank winged it with a shotgun blast, and had kept it alive while its wound healed. Unfortunately the damage was such that it would never fly again so I felt compelled to bring it home with me. The fact that I also wanted to be the possessor of the most unusual pet any of my peers had ever seen may have had something to do with this decision.

The jaegar *was* unusual — unusually unpleasant. It would strike at any living thing that came within reach, and its hooked beak could rend flesh and draw blood with ease. It was handsome, but its manners were foul. It defecated as if equipped with an anal blunderbuss and woe betide anyone who stood behind it. Furthermore, it had a terrible voice — a kind of whining scream with which it gave almost continuous expression to its animosity.

Uncle Frank and Bert climbed back aboard the train, perhaps with some relief. Angus and I waved them off, then piled my gear and trophies into Eardlie and drove out to the country club where the caravan, my mother, Mutt, and Wol awaited us.

Wol had nearly attained his full growth during my absence. He now stood two feet tall and, although he had not yet taken to flying, had a wing spread of four feet or more. He had also acquired a regal presence. I think he saw the rest of animate creation, humans included, as his subjects. Certainly he tended to treat them as such. For the most part he was a benign monarch, unless roused by an act of *lèse-majesté* when he could become a Jovian thunderbolt. He admitted me to his court by leaping lightly on my shoulder, breathing deeply in my ear, and muttering a gruff "Hoo-hoo-hoo."

Mutt crawled out from under the caravan and flung himself upon me *after* Wol had returned to his throne, a poplar post beside the fireplace.

"I think Mutt's trying to tell you," Angus explained, "that he's had enough. Wol sneaks up on him when he's asleep and grabs his tail with those ruddy great claws. Thinks it's a great joke. But it's turning Mutt into a troglodyte."

My mother kissed me warmly. "You've gotten thinner, Bunje darling, though as smelly as ever. How about a swim? That is, if you can make your owl stay away from us. He's taken to hopping on people's heads when they wade in and if you try to shake him off he gets excited and goes to the bathroom in his pants." She shuddered delicately.

It was apparent that Wol was pushing everyone's tolerance to the limit. I suspected my parents were only waiting until I had settled in to tell me the time had come for Wol to return to his own world.

If Wol was difficult to live with, the jaegar turned out to be hell on wheels. It absolutely refused to accept life in a cage and raised such a hideous protest that my mother began to fret. "What on earth will people at the club think we are doing down here?" she asked. "It sounds like murder!" Murder would have been a solution but we could not bring ourselves to that so we tried giving the jaegar its freedom. It abused that freedom with diabolical ingenuity.

Unable to fly, it developed the agility of a road-runner. It would lurk behind a tree, under a bush, or beneath the caravan, then burst out unexpectedly to viciously assault a bare leg (if human) or a hairy one (if Mutt). One ripping tear and it was gone, far too swiftly for any of us to catch it. It became the chief peril of our camp. Then one day it spread its reign of terror farther afield. Bruce, who was working as a caddy at the club, came down to tell me: "Your goddamn bird tore into a ladies' foursome on the fifth hole this morning. Jeeez! You should have seen 'em scatter! The pro is frothing mad and says he'll shoot the damn thing next time he sees it."

Since there was nothing else to be done I caught the jaegar (by trapping it with meat scraps) and caged it again. By the following day its screams of rage had become unendurable, so Angus and I built a good-sized coop of chicken wire, reasóning that the bird might be more content in a larger enclosure.

Nothing of the sort. It raced around the perimeter of the coop, tearing at the wire and making such a fuss that, belatedly, the King at last deigned to take notice of it.

None of us actually saw what happened. I think Wol hopped over the wire impelled by simple curiosity about the nature of this obnoxious bird. And I think the jaegar made the mistake of going for him.

The jaegar's two long tail feathers became a souvenir pinned on my bedroom wall at home. Wol became a hero. And the suggestion that the time had come for him to return to the wild was never voiced.

We continued in our camp by the river through most of the remainder of the summer. Angus commuted to work in town, though no longer by automobile. Late in 1935 he had bought a sailing canoe which had somehow become stranded in Saskatoon. It had fallen on hard times but Angus spent much of the winter lovingly restoring it. He painted it green and christened it *Concepçion* in honour of the event with which my life began.

Each summer morning he slid *Concepçion* into the river and, if the wind would serve, set sail for Saskatoon six miles downstream. Off he would go, skimming the shoals and slipping past the dead-heads until he reached his chosen landing place below the Bessborough Hotel. Then, having unstepped the mast and furled the sail, he would flip the canoe upside-down on his shoulders and portage it through the morning traffic to the library. In the evening he would reverse the procedure. If he could not sail he paddled, although it could take as long as three hours to struggle home against the current.

I sometimes accompanied him to town but the attention he attracted as he trotted down Broadway with a sixteen-foot green canoe over his head was not something I relished. One morning as we sprinted to catch a traffic light, we encountered a school acquaintance of mine. He watched in astonishment as my father sped by then turned to me, bringing up the rear, and exclaimed, "Holy Cow, Billy! Your old man sure is one loony bird!" I could only nod miserably and hasten on.

We had to strike camp and return to the city earlier than we wished in order to move to a different house before school started. The new one was on Spadina Crescent on the opposite side of the river. The river bank here was much lower and less interesting, but our new home was surrounded by unkempt grounds and bordered by rows of big poplars which attracted birds and small animals. The house was, according to Helen's diary, "our best yet, with five bedrooms and two fireplaces, and very cosy."

A third floor, consisting of two bedrooms, comprised what was effectively a separate apartment. To my delight I was given both rooms, one to sleep in and the other to use as "your study, Farley, although," added my father, "God alone knows what you'll really do up there."

Predictably, one of the things I did was populate the third floor with a variety of animals, dead and alive. I kept a dead badger in my study for a week while I tried to skin and stuff it, and if God didn't know about *that* He was the only one who didn't.

Some of my other acquisitions were thrillingly alive. One such was a prairie rattlesnake captured by a friend during his summer vacation at Val Marie in the south of the province. He parted with his still-youthful snake — it was barely three feet long — reluctantly and only because his parents adamantly refused it house room. Although I was confident that *my* parents would be more understanding, I did not put the matter to the test by telling them about the handsome guest who lived in the bottom drawer of an old bureau in my study.

Kaa, as I called him after the rock python in the Mowgli stories, was a gentle soul who did not object to being handled. He was usually so full of mice that probably all he wanted to do was sleep. I kept a colony of white mice in the drawer directly above Kaa, which made for a most convenient arrangement.

The study was festooned with birds' nests and other natural bric-à-brac, including the wings and tails of hawks, owls, and an eagle, all of which had been shot by sportkillers as vermin. My formidable collection of books, mostly concerned with natural history, was ranged around two walls. A table under a dormer window in the third wall supported an aquarium and a small microscope. A desk and chair occupied that portion of the floor not already crowded with cardboard cartons and home-made cages.

The Churchill experience coloured the remainder of that year. I spent most of my evenings in my eyrie (Helen's name for it) dreaming over my summer's notes and specimens, poring over Arctic maps with future expeditions in mind, and reading everything available about the North and its creatures.

All the same, I did not neglect the world around me. Murray, Bruce, and I remained boon companions and we trekked ever farther afield by bicycle, snowshoes, and on foot. A tribe of three, we frequently camped out on long weekends and other holidays.

I spent part of my time that autumn banding small migrant birds which I caught in a variety of wire-mesh traps set in our back yard. By mid-October I was catching, banding, and releasing as many as a hundred white-throated, white-crowned, song, and Harris sparrows a week. And this despite

Wol's attempts at sabotage. He had cast himself in the role of an avian Scarlet Pimpernel. He would sidle up to a trap in which a dozen small birds were fluttering and rake it with the talons of one foot, tearing a wide enough gap in the mesh to allow the captives to escape. Oddly enough he never tried to snatch a quick snack for himself. He would stand aside with a distinct air of benevolence and watch the prisoners flee. I dealt with this problem by whacking him with a rolled-up newspaper whenever I found him near one of my traps.*

Although I was an enthusiastic bird bander, my work was not entirely appreciated by the Ottawa civil servants in charge of the program. A few years ago a well-wisher sent me the file of my 1936 correspondence with the National Parks Branch. Pinned to the cover of the file was this somewhat querulous note:

"This [banding] co-operator has been the cause of considerable confusion in this Department because he seldom uses the same name twice. Here is a list of the names he uses:

W. F. and W. Farley Mowat

William Mowat

Billy M. Mowat

Farley Mowat

William Farley Mowat

Farley McGill Mowat

William McGill Mowat

---

*Wol and a second horned owl named Weeps, whom I acquired in September, played a major part in my life during this period. But I have already told that part of the story in Owls in the Family and The Dog Who Wouldn't Be.

"And sometimes he does not sign his name at all."

I don't recall having had that much of an identity problem but I had so much on my mind that sometimes I may have been a little uncertain about who I was.

As if having to make the adjustment from Arctic adventurer to high-school student was not enough to unsettle me, my hormones had begun to give me hell. Wet dreams and masturbation no longer sufficed to ease my distress. I was becoming ever more desperate to experience the real thing.

While in the throes of this turmoil I made a new friend. Munro was the son of a near neighbour. Although my elder by only a year in time, he was several years older in experience. Tall and loose-limbed, with lanky hair and a blond fuzz on his upper lip, he looked to be at least nineteen — and was a whiz with girls.

Not only did he squire them to dances, he parked with them in his father's car in secluded spots along the river bank and, although he never actually told me so, I had the distinct impression that he had been "all the way" with several of them.

I hung around with him as much as I could, hoping he would help *me* find *my* way to paradise. The basic difficulty confronting me remained what it had been for years. I continued to look like everybody's kid brother, and to be treated like background music as far as sex was concerned. I simply could not get a girl to take me seriously as a suitor, or even as a threat.

Munro did his best. He arranged a blind date for me with one of his girlfriend Myra's chums. The plan was for us to take the girls to a movie, buy them a soda, then drive to a secluded spot near the ski jump where we could neck and — maybe — *do* it!

In a bold if misguided attempt to make me appear older, Munro borrowed his mother's make-up. He darkened my eyebrows, gave me the faintest hint of five-o'clock shadow (together with an incipient moustache), and painted what were supposed to be bags resulting from a dissolute life under my eyes. The result was to turn me into a caricature of a juvenile Dracula but I was desperate enough to try anything. I listened attentively to Munro's advice.

"You've got the gift of the gab if nothing else, Billy. So what you gotta do is *talk* Violet into it!"

It was pitch-dark on a bitterly cold October night when we reached the parking spot. Munro and his girl wasted no time steaming up the front windows while I tried to make out in the back seat.

It was no dice. No matter what brilliant verbal sallies I employed, they got me nowhere with dumpy little Violet, who was nearly seventeen and knew her way around. She listened to my inspired pleadings with gum-chewing imperturbability until her patience was exhausted. Then she leaned over the back of the front seat and loudly demanded of Munro, "Don't this here kid *ever* shut *up*?"

This so devastated me that I got out of the car and walked home.

Munro commiserated with me the next day. "Maybe you should have used your hands more and your mouth less," he mused. "But Violet's probably too old for you anyways. How'd you like to try again with Myra's young sister?"

"How young?" I asked suspiciously.

"Well, she's twelve going on thirteen."

Celibacy seemed preferable, and safer.

At this juncture I concluded that, if nature was not going to bestow upon me a patina of physical maturity, I would have to do it for myself. I began leading the dissolute life Munro had tried to counterfeit using his mother's cosmetics.

I made a point of staying awake as late as possible every night until I really did develop such dark shadows under my eyes that Helen became alarmed and took to giving me huge glasses of Ovaltine to build me up. I rather liked the stuff but used to wonder what my mother's reaction would have been to the Ovaltine song, which was one of our school-yard jingles.

> *Uncle George and Auntie Mable*
> *Fainted at the breakfast table.*
> *Ovaltine soon put them right;*
> *Now they do it noon and night.*

Smoking was another dissolute activity in which I engaged. Like every kid in Saskatoon, I had experimented with bits of burning reed or punk, then with the occasional, usually shared cigarette. Now I took up the vice with dedication. I snitched one of Angus's old pipes, stuffed it with Pic-o-bac tobacco (the cheapest available) and went around wreathed in blue smoke whenever I was out of sight of my parents or other censorious adults.

And I took up booze.

This was not easily done because we were still living under the bleak memory of Prohibition, and the sale of alcohol was tightly controlled. Fortunately, my parents and their circle of friends liked to drink and did so on all possible occasions. I used to make a point of volunteering to help clean up after parties at our house. Having carefully collected all the glasses, I would sneakily pour their residues into an empty bottle hidden in the broom closet. It was surprising how much good stuff people left undrunk. Sometimes I collected as much as a pint of what could be quite a potent concoction.

Sipped in the privacy of my study, the lees made me woozy, drowsy, and sometimes sick to my stomach. There was no pleasure in it. Eventually I concluded I was going at it the wrong way. Instead of tippling, perhaps I should try binge drinking. With this in mind, I saved about a quart of party juice in a screw-top molasses jug to which I added a pound of brown sugar, some apple cider, and a cake of yeast. Then I stored the jug in a niche under the cellar steps and temporarily forgot about it.

Fortunately I was alone in the house when it reminded me of its existence. One Sunday afternoon I heard a smothered "blooop" and shortly thereafter the house began to stink like a combination brewery and distillery. I followed my nose down the cellar stairs to find the air thick with fumes and nothing remaining of my jug except shards of pottery and a noisome puddle on the concrete floor.

Since making moonshine was a criminal offence, I was afraid my father would not treat this mishap as a mere peccadillo. This time, I feared I was in deep, deep trouble.

Frantic, I mopped the liquid residue into a pail, carried it outside, and poured it down the river bank. However, even with all the doors and windows open to the chill November air, I could not dispel the tell-tale stench that filled the house. My fate seemed certain until I remembered that down in the cellar was something which might prove to be my salvation.

The doors and windows still gaped wide when my parents arrived home an hour later. I met them with an apologetic smile.

"What the devil's going on?" Angus demanded. "Are you trying to heat the whole of Saskatoon?" He paused, sniffed, and then looked startled.

"What's that terrible smell? You *haven't* been fooling around with chemistry experiments again?"

No, no, I reassured him. I was very sorry but I had accidentally broken the gallon glass jug of Fly Tox he kept in the cellar and it had spread all over and, of course, the fumes had filled the house. I was really, *really* sorry.

The house stank of Fly Tox for weeks, but every now and then I caught a whiff of a different though equally pungent aroma. Fortunately, my parents were not as gifted in an olfactory way as was their son who, I may say, had now — temporarily — renounced most of his dissolute ways.

EIGHTEEN

The drudgery imposed by school that winter was relieved by one brief yet shining adventure. My parents had become friends with painter Ernst Lindner and his beautiful wife Boadil. The four of them agreed to make a mid-winter sortie into the "bush" north of Prince Albert where Ernie knew some old-style trappers. I was to go too and, when I pleaded for the company of Bruce Billings, he was also invited.

We left Saskatoon on December 28, travelling in style. "Jamie" Jamieson, another neighbour on Spadina Crescent and head of the Canadian National Railways system in Saskatchewan, had made his private rail car available to us. Although of much the same vintage, it was a far cry from the old colonist car that had trundled me to Churchill. Jamie's was fit to carry royalty and had done so on at least one occasion. It had a spacious lounge stuffed with mahogany and walnut furniture; two en-suite bedrooms; a dining room complete with gilt chairs and rosewood table; its own galley, chef, and steward.

None of us had ever experienced such luxury before. That evening our elders dined on rare roast beef and drank French wines while Bruce and I were entertained by the chef, who fed us half a fried chicken each washed down with hot chocolate served in silver mugs.

The contrast with what awaited us at Prince Albert next morning was stunning. For one thing, the temperature when we disembarked from our palace on wheels was 40° below. For another, the transport which was to carry us north to a logging camp on the next leg of our journey turned out to be an antiquated open truck. It had just brought several horses south and the only preparation the driver had made for us was to spread fresh straw on top of the frozen turds. We climbed on top of the straw and huddled together under our sleeping bags and blankets. A suggestion that our women might ride in the battered cab was met with a shake of the head from the driver.

"Can't do that. Cab's full o' stuff that'd freeze solid if it was in back."

Never mind. It was all adventure; besides which, Ernie and Angus had been provided with a thermos jug full of hot rum toddy by Jamieson's thoughtful steward. Even Bruce and I were allowed the occasional swig, if only to ensure that *we* did not freeze solid.

After some hours of bouncing along an ice-rutted road,

we reached the logging camp where three teams of dogs hitched to canvas-sided carioles awaited. Shortly thereafter we were being driven across a frozen lake then through miles of forest trails to our final destination, a cluster of log cabins housing the family of trappers who were to be our hosts.

A cabin had been prepared for us by clearing out the wolf and fox skins which had been stored in it. The aroma lingered on — a wild, sharp odour not unlike that of the circus menagerie that had visited Saskatoon. There were bunk beds along three walls; a sturdy cook stove; and a red-hot, pot-bellied heater in the middle of the room. Although the temperature outside was close to fifty below we were comfortable enough. But when I awoke next morning it was to find the hair of my head cemented to the log wall with hoar-frost.

We spent a week in the bush, during which Bruce and I snowshoed behind the dogs, accompanying our hosts to traplines that ran many miles from camp. It was so cold that we saw few live animals — an occasional raven, a fox, and a few flocks of finches. However we saw enough dead and dying ones. I especially remember a young wolf caught by a snare around its neck. For some unknown length of time it had been slowly choking to death. We heard it before we came in sight of it — a horrid, gurgling, rasping sound that haunts my memory still. At another "set" we found the paw and foreleg of a fox in a sprung trap. The animal had chewed its own leg off in order to escape.

I cannot claim that I was opposed to trapping then. In truth, I greatly admired and liked the trappers and envied them their way of life. Yet in retrospect, I realize that the animal victims I saw that week were probably the origins of the passionate aversion I now feel for the fur trade in all its aspects.

The early months of 1937 were haunted by Arctic dreams, which seemed appropriate since we were experiencing an especially bitter winter with frequent blizzards and weeks of below-zero temperatures. But I hardly noticed the weather. Shortly after Christmas Uncle Frank had written to compliment me on the work I had done for him at Churchill and to ask if I would like to go north again. This time, he said, we *would* reach Seal River and might even voyage north along the Hudson Bay coast accompanying "Windy" Smith to the legendary Thlewiaza River. "Windy" had told Frank of an unknown species of freshwater seal living at Edehon Lake upstream on the Thlewiaza, where several Inuit families made their summer camps, and which was virtually unknown country to any whites except himself.

My parents were again agreeable. However this time they attached a caveat. I could go if, and only if, my school work improved dramatically. This sort of blackmail has pushed a good many students through school. It worked on me. I became a model student — and hated every minute of it.

In the end it was all to no avail. Come March Angus was offered and quickly accepted the post of Inspector of Public Libraries for the Province of Ontario. He was to begin his new job on July 1, which meant that we would have to move back east — lock, stock, and barrel — early in June. This in turn meant the dissolution of my Arctic dreams, or at least their indefinite postponement. In the event, I was not to visit the Thlewiaza and Edehon, or meet the Inuit and the freshwater seals until ten years later — but that is another story.

My disappointment was acute and devastating. Not only was I going to be deprived of the Arctic experience, I was going to be deported from the one place I had ever really felt at home. I became depressed and, in classic teenage style, grew sullen and uncommunicative. Helen noted in her diary: "This has changed poor Bunje into a perfectly horrid boy." Nevertheless she was sympathetic, whereas my father was not.

"It'll be the best thing that ever happened to you. Going back to where you came from will do you the world of good. Chin up and take it like a soldier and a man!"

But I was neither. In the privacy of my third-floor haven, I alternately raged and wept. I made wild plans to run off and become a cave-dwelling hermit on the banks of the Saskatchewan where I would be supplied with necessities by my pals. Bruce even volunteered to run away with me.

"Let's the both of us light out for the bush. We can be trappers and live with the Crees up around Lac La Ronge. With Mutt and Rex, we already got the best part of a dog team…"

My human friends did their best but in the end it was not they who calmed my anger and eased my misery. It was the

Others. Spring returned to the prairie and that great cycle of renewal again drew me into its vortex — drew me in, solaced me, and led me to an inchoate recognition of a sustaining truth: that, no matter where I might find myself on the face of the planet, I would never be out of place so long as the Others were there to comfort me with their presences.

My sixteenth birthday was approaching and one day my mother asked how I would like to celebrate it.

I knew exactly what I wanted.

"I'd like to go away for a week to Dundurn and camp out there by the big slough with Murray and Bruce. Just the three of us on our own, and Mutt and Rex, of course."

She smiled. "I'll talk to your father about it, Bunje dear. He's already feeling a teeny bit guilty about taking you away from all this." She waved her hand as if to encompass the whole of the Saskatchewan plains. "Perhaps I can make him feel a little more so."

She was as good as her word.

Early on the morning of May 12 we three boys loaded ourselves, dogs, and gear into Eardlie, and Angus drove us to the poplar bluff south of the hamlet of Dundurn which I had chosen as our campsite. He did not linger and I did not encourage him to do so. It would be some time before I would forgive him for so abruptly changing the tenor of my life.

Perhaps in compensation for a dreadfully hard winter, spring had come early. On my birthday the sky was bell-clear and the sun beat down so brilliantly that the reflections from the wavelets on the big slough hurt the eyes. The new grass was vividly, lusciously green and the whole world smelled richly musky and alive. The floor of the bluff was coated two or three inches deep in silky "snow" from the cottonwoods. It kept getting in Mutt's nose and making him sneeze as he sniffed around after chipmunks.

Murray, Bruce, and I felt so good we began playing like kids. We lay down on our backs and waved our arms, trying to make angels in the fluff. Then we gathered piles of the silky stuff to put under our bed rolls, but it was so evanescent that we ended up sleeping on hard ground.

Once the tent was pitched and a fireplace cleared, we went off to visit the slough. It was about two miles long and half a mile wide. Although it had been mostly dry in recent years, this spring it was overflowing and water stood knee-deep amongst the cat-tails around the shores. But it wasn't just full of water; it was overflowing with life as well.

Out in the middle, two or three hundred whistling swans formed a flock so dense it seemed like a great ice floe. We stopped at the edge of the rushes to stare in wonder at the multitudes of birds crowding the surface of the slough.

"Holy cats!" said Murray. "There must be a million ducks and geese out there...but what the heck are *those*?"

He pointed to the southern sky out of which a squadron of airborne creatures was descending with such stately and primaeval mien they might have been pterodactyls coming to us out of Jurassic times. They were white pelicans — enormous, ancient fliers whose antiquity does indeed go back to the age of the dinosaurs. A wild flurry of ducks scattered out of the way as the pelicans planed down to strike the surface with a spray-drenched whoosh.

We began working our way around the circumference of the slough. A stretch of muddy foreshore where cattle had trampled down the rushes had become a living carpet of shorebirds: marbled godwits, long-billed curlews, and a dozen magnificent cream, russet, and black avocets, mingling with hordes of smaller species.

Heedless of hunger we continued to follow the shore, seeing waterfowl of so many varieties and in such abundance that I gave up trying to keep a record of them. Our ears were full of the rush of wings and the cacophony of thousands of voices gabbling about sex, food, travel, and whatever else birds talk about.

Led by Mutt and Rex we waded through acres of reeds and rushes, constantly assailed by belligerent red-winged and yellow-headed blackbirds claiming nesting territories. Muskrat houses freshly crowned with steaming layers of swamp muck stood like miniature islands amongst the reeds, and a pair of coots or grebes seemed to be nesting on top of every one. Marsh wrens were weaving their hanging nests on cat-tail stems, and unseen sora rails yammered at us for trespassing on their watery turf.

Tired out we at last straggled homeward, but had to halt while Rex excavated some hillocks of black mud heaped up by pocket gophers. He uncovered none of the secretive mammals themselves, but did reveal hundreds of big tiger salamanders, eight and nine inches long, who were using the burrows as way-stations on their journey to the slough to spawn.

It was late afternoon by the time we flung ourselves down beside our tent. The emerald-leafing poplars overhead were alive with waves of warblers and other small fry avidly competing for the first emerging insects. Interspersed among them was the azure blaze of mountain bluebirds, the orange challenge of orioles, and the flame of rose-breasted grosbeaks.

By then we were too tired and too surfeited with this abundance of living things to pay much attention. When a long skein of sandhill cranes flew low overhead, trumpeting their sonorous refrain, I did look up, but only briefly. Then I lay back in the cotton snow and closed my eyes.

"Hey, Brucie," I said sleepily. "We've sure seen an eyeful today. You think we'll ever see anything like it again?"

Bruce was sucking on a straw. He took it out and threw it away.

"Maybe, yeah. I guess so, if we're lucky."

We were not to be that lucky. I doubt if anyone else ever will be either. I think it is too late.

Six days later Eardlie came puttering up to our campsite. The tent was already down and everything was packed. We loaded the gear aboard the little car.

And it was here, at this time and in this place, that I really said goodbye to the prairies; to Bruce and Murray; to Mutt and Wol; and to all the Others with whom I had lived the happiest and, it may be, the best years of my life.